Mark Walker's career has been centred on the use of genetic engineering techniques to improve human health, particularly by the development of safe and efficacious vaccines. He is currently Associate Professor in Molecular Genetics at the University of Wollongong's Department of Biological Sciences, one of the leading teaching and research departments for biological sciences in Australia.

David McKay has a strong interest in both research and education and has taught science at both high school and tertiary levels. He currently lectures in biochemistry, microbiology and biotechnology at the University of New England, Australia and recently received the Vice-Chancellor's award for excellence in teaching at that institution. David's research interests include biomediation and environmental microbiology.

Unravelling Genes

A layperson's guide to genetic engineering

Mark Walker
David McKay

ALLEN & UNWIN

First published in 2000 by
Allen & Unwin
9 Atchison Street
St Leonards NSW 1590
Australia
Phone: (61 2) 8425 0100
Fax: (61 2) 9906 2218
E-mail: frontdesk@allen-unwin.com.au
Web: http://www.allen-unwin.com.au

National Library of Australia
Cataloguing-in-Publication entry:

Walker, Mark Joseph.
   Unravelling genes: a layperson's guide to genetic engineering.

   Includes index.
   ISBN 1 86508 086 1.

   1. Genetic engineering. 2. Genetics. 3. Biotechnology.
   4. DNA. I. McKay, David B. II. Title.

660.65

Set in 11/12.5 pt Plantin Light by DOCUPRO, Sydney
Printed by KHL Printing Co. Pte Ltd, Singapore

10 9 8 7 6 5 4 3 2 1

Contents

Figures and tables

Figures

Tables

Acknowledgements

The authors would like to thank Manfred Rohde and Alan Adolfsson for photography and Bio-Rad Australia for supplying Figure 4.1. We would also like to thank John Pemberton for supplying us with the violacein genes. Thank you to Ted Steele for encouraging us to write this book and to our families for their support.

Preface

The aim of this book is to give you an impression and understanding of what genetic engineering is, what it can do and, just as importantly, what it can't. To achieve this aim we've written this book with the philosophy that science, although not always easy, is approachable, relevant and understandable to the inquiring layperson. In line with this philosophy, we have tried not only to present the ideas involved in genetic engineering, but also to make these ideas relevant by presenting you with everyday analogies and even some experiments you can do in the kitchen.

How is this book arranged?

Genetic engineering (as a field of study) is a part of biology, and it is the principles of biology which are the foundation of genetic engineering technology. This means that to understand something about genetic engineering, you first need to understand some basic ideas about biology. (Just as, if you wanted to learn something about how the electric lights in your home work, you would first need to understand something about the theory of electromagnetism.) So in the first part of this book, we look not at genetic engineering but at the main biological principles which underpin it.

The body of this book deals with the many and varied technologies of genetic engineering. We have divided this section according to the type of organisms used, starting with the simplest systems, the bacteria. We then examine the use of plants, animal cells and animals as factories for the manufacture of genetically engineered products. We will explain why cloning animals ('Dolly the sheep') is an important technological advance in the field of genetic engineering, and finally we shall examine methods by which gene therapy can be carried out in humans.

No technology exists in a vacuum, and we feel that to completely ignore the implications of a new technology from an ethical standpoint is a dangerous exercise. However, whilst ethical issues cannot be ignored, a full treatment of them is certainly beyond the scope of a book of this nature. We have thus touched on some illustrative examples of ethical dilemmas arising from genetic engineering, but our main focus is to give you a clear idea about what genetic engineering technology is, as simply and impartially as possible. Armed with this understanding, you should be able to develop informed opinions about the repercussions of employing this technology.

1 Towards an understanding of genetic engineering

Although you really need to read this whole book to get a good impression of what genetic engineering is, we thought we'd start by giving you a rather brief definition and thumbnail sketch of what this technology is all about.

Genetic engineering (also known as gene cloning or recombinant DNA technology) involves alterations of an organism's genetic material which is made of a chemical called DNA (deoxyribonucleic acid). Either the existing genetic material of the organism is changed (for example, by cutting some out) or, more frequently, by taking DNA from one organism and putting it into another.

Why then has there been such a fuss about this technology? Why does genetic engineering have the potential to revolutionise all our lives in the twenty-first century with the same impact, for instance, that the development of electricity had on the lives of our grandparents?

There are probably many reasons why people have anxieties about genetic engineering. One reason is simply the fact that it's such a new technology and it is developing rapidly. There is a 'shock of the new' or a 'future shock', to borrow phrases from Robert Hughes and Alvin Toffler. Future shock is probably a more apt term, meaning that the shock comes not so much from the change *per se*, but from the *rate* of change. People generally find it difficult to keep pace with new technological developments and genetic engineering certainly fits this category; we, the authors, find keeping pace with new developments in genetic engineering difficult, and we work in the field!

However, it is worthwhile keeping in mind that the antagonism towards technologies that appear suddenly and develop rapidly is not new. A case in point would be the development of powered flight. A knee-jerk reaction at the inception of this technology was the statement, 'If men were supposed to fly they would have been given wings', which typifies an attitude, not of rationality, but of a primordial reaction to things new, not understood and therefore to be feared. Similarly, a typical catchcry today would be that 'genetic engineering is not natural' or 'we are playing with nature'. Such arguments are, to use the Australian vernacular, cop-outs; they are irrational and an easy way out. Isn't it much easier for closed-minded people not to put in the hard work to try and understand the complexities of the technology, its limitations and benefits? Why not make a few glib, 30-second comments—the kind of thing politicians are adept at?

We agree that genetic engineering is playing with nature, that it is not 'natural'. But how can we possibly use this as a rational argument given that all of our technology is to some extent unnatural and is playing with nature? Granted, there are many things that we, the authors, feel should not be done in the field of genetic engineering, but our reasons for feeling this way are complex. Given what we have just said, we hope you can better appreciate our intentions in this book—a thorough analysis of genetic engineering given constraints of time

Figure 1.1 Chronology of advances in gene technology, compared to the development of powered flight

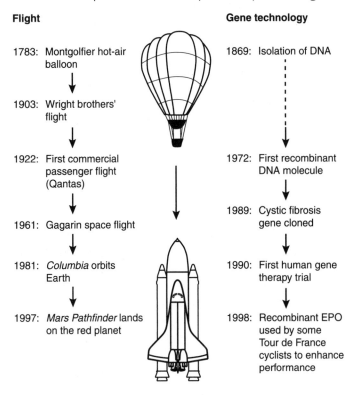

Flight

1783: Montgolfier hot-air balloon

1903: Wright brothers' flight

1922: First commercial passenger flight (Qantas)

1961: Gagarin space flight

1981: *Columbia* orbits Earth

1997: *Mars Pathfinder* lands on the red planet

Gene technology

1869: Isolation of DNA

1972: First recombinant DNA molecule

1989: Cystic fibrosis gene cloned

1990: First human gene therapy trial

1998: Recombinant EPO used by some Tour de France cyclists to enhance performance

and the complexity of the topic, so that you can form rational judgments and opinions based on facts. No glib comments, no 30-second sound-bites.

For you to appreciate just how new genetic engineering is, and what a rapid impact it has made on technology and biology, we thought it worthwhile to compare the time frame for the development of powered flight with genetic engineering (Figure 1.1). You can see here that genetic engineering is only 30 years old. It is little wonder that the key concepts of genetic engineering are only now starting to be taught at high school.

Genetic engineering is a technology whose roots run deep in many areas of biology. It was a basic knowledge and understanding of biology which gave rise to genetic engineering, and thus it is not surprising that to really understand the technology of genetic engineering, you must understand at least some principles of biology.

So in order to understand more about what our more simple definition of genetic engineering ('taking a piece of DNA from one organism and putting it into another organism') means, you will need at least a working knowledge of what organisms, cells and DNA are.

Cells and organisms

On April 13, way back in 1663, Robert Hook first demonstrated the microscope to the Royal Society in London. When he placed a slice of cork (plant tissue) under the microscope, he made the following observation: 'Our microscope informs us that the substance of cork is altogether filled with air and the air is perfectly enclosed in little boxes or cells distinct from one another'. It soon became clear that all organisms are made of functional units called cells.

Cells are the common denominator, the functional unit, of all life on Earth. They have the unique ability to take up chemicals from the environment (feed themselves), eliminate waste products, self-replicate and grow. No other things on the planet have all these unique abilities. Interestingly, there is a natural limit to how large cells can grow and, because of this, larger living things are built not by increasing the size of their cells, but by increasing the number of cells. (Elephant cells are the same size as those of a mouse—generally around two-hundredths of a millimetre in diameter.)

All living things may be categorised as 'organisms'. Humans, trees, fish, birds, fungi, algae and the 'bugs' that cause diseases such as tetanus, diphtheria and whooping cough are all classified as living organisms. Organisms can be further classified or grouped together in several ways. Organisms can be classified into groups such as: multi-cellular organisms, that is, humans, trees, fish, birds, fungi and multicellular algae; and unicellular organisms, that is, certain types of single-cell algae and many of the 'bugs' or bacteria that cause disease. This is simply a size classification whereby large organisms are made up of many cells, that is, they are multicellular.

An average-sized human is made up of approximately five trillion cells (that's five times a million times a million). In fact, any living organism that we are able to see with the naked eye will usually be a multicellular organism. On the other hand, unicellular organisms cannot normally be seen without the aid of a microscope. The organism frequently used in genetic engineering projects, *Escherichia coli* (or *E. coli* for short), is a unicellular microorganism.

Organisms, or the cells they contain, can also be classified in another way, according to the structure of the cells. The two divisions in this classification are eukaryote and prokaryote. Biologists find these divisions very useful, but you may not have heard about them before. The names give it away if you happen to be a Greek scholar—'eu' means 'true', 'pro' means 'before' and 'karyon' means 'nucleus'. In other words, a eukaryote has a true nucleus whereas a prokaryote doesn't. We'll actually look at some pictures of prokaryotic and eukaryotic cells in a moment to illustrate the difference, but for now the take-home message is that whereas the multicellular and unicellular classification really only tells us something about the size and cellular organisation of an organism, the eukaryotic and prokaryotic classification system tells us something about the structure of cells in an organism. As it happens, all multicellular organisms are made of eukaryotic cells, whereas a unicellular organism consists of either a eukaryotic or a prokaryotic cell.

You're probably now asking, 'What sort of organisms are prokaryotes and eukaryotes?' The best way to answer this is to give some examples of prokaryotes and eukaryotes.

Prokaryotes are made up of two groups, called the bacteria and archaea. Examples of bacteria are: *Mycobacterium tuberculosis* (the organism which causes tuberculosis), *Neisseria gonorrhoeae* (which causes gonorrhoea) and *Acetobacter* species (used in the production of vinegar). The archaea are a strange bunch of prokaryotes which live in unbelievably harsh environments. An example is *Sulfolobus*, which lives in sulfur-rich hot acid springs at temperatures up to 95°C!

Eukaryotes are all the multicellular organisms such as plants and animals. Unicellular organisms such as yeasts are also eukaryotes.

Prokaryotic and eukaryotic cells

Eukaryotic cells, such as the cells that make up our bodies and those of all multicellular organisms, are much more complex than those of a prokaryotic organism such as *E. coli*. However, despite the tremendous differences between prokaryotes and eukaryotes, there are still a lot of similarities between them. For instance, both types of cells are bound by a cell membrane. As an analogy for a cell, imagine a plastic bag filled with water: the plastic bag represents the cell membrane and the water represents the fluid contents of the cell. However, cells, and in particular eukaryotic cells, are much more complex than a fluid-filled sack of water. Eukaryotic cells, particularly from a structural point of view, are exquisitely complex.

Figure 1.2 Electron microscope sections of a prokaryotic cell and a eukaryotic cell

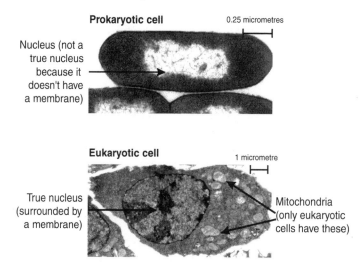

Using specialised microscopes called electron microscopes, it is possible to see the contents of both prokaryotic and eukaryotic cells. In the same way that it is possible to examine the core of an apple by slicing it in half with a knife, the cells in Figure 1.2 have been sliced in half with a diamond knife to reveal their contents.

Notice in this figure the presence of many organelles—separate structures which perform different functions in the eukaryotic cell. The mitochondria, for example, are a type of organelle that is responsible for energy production within the cell. Each organelle is bounded by a membrane serving to divide the cell into separate compartments. Also notice that the eukaryotic cell has a true nucleus, meaning it is bounded by a membrane, whereas the prokaryotic cell does not.

Whilst the structural differences between prokaryotic and eukaryotic cells are great, you should also bear in mind that from a chemical standpoint, there are a vast number of similarities as well. All cells, be they prokaryotic or eukaryotic, contain DNA as their genetic material. All cells contain proteins and membranes. It is because of these similarities that we can often regard all cells as functioning in the same way.

In the later chapters in this book, you will often encounter the idea that all cells operate and function in a similar way and that you can disregard the differences between prokaryotes and eukaryotes. Sometimes though, these differences cannot be ignored and can have a large influence on whether a particular genetic engineering project will work.

Multicellular organisms and the division of labour

As you are sitting here reading this book, consider the fact that your body is composed of five trillion individual cells. Although they are all cells, and have many things in common, these cells are also different and perform different functions. For example, your muscle cells, which enable you to turn the page, convey movement whereas your brain cells, which allow you to understand the words, convey electrical impulses. From a cellular perspective, how are complex, multicellular organisms arranged?

It may be helpful to consider an analogy here. Think of a single cell in your body as being analogous to a house in a large metropolis. The house has a floor, walls and roof, which are analogous to the membrane of a cell. The house is not isolated from the rest of the city, and there are many forms of communication, such as roads, train lines, television, radio, newspapers and mail serving as lines of communication to all other houses and their inhabitants. Similarly, a cell is not isolated from other cells of your body. Cells are connected to each other via the blood system and lymphatic system. Cells also have many different antennae on the cell surface, called 'receptors', each of which is responsible for receiving one of the different types of 'messages' (in the form of different types of chemicals) that are constantly being sent around the body. Cells have a number of doors and windows (called pores, vacuoles and pinocytotic bodies) whereby foodstuffs and the garbage can enter and exit. When food is taken into the house, it is normally taken to the kitchen. After entry into the cell, food is normally taken to organelles of the cell called mitochondria, where it is broken down to produce energy-containing molecules called adenosine triphosphate (ATP). ATP is then used by other parts of the cell as an energy source to perform work. Without ATP, none of the other 'appliances' (organelles) of the cell would function. Amazingly, some plant cells have a light-harvesting appliance called a chloroplast—which would be analogous to a house with a solar panel.

In our house, there is a computer which automatically controls all the appliances and smoothly runs the house. This computer contains all the information and plans for the construction of all the houses in the city including this house, the manufacture of household appliances, maintenance instructions when things break down, which type of nails to be used for nailing the roof on, the colour of the walls, what pictures are hung on the walls and even the recipes used in the kitchen and the type of dishwashing powder used for the dishwasher. Exact copies of this computer actually sit in every single house in the city. Although all the computers in all the houses contain all this information, each computer only ever uses a small amount of the information, for instance by switching on and off appliances only in its own house.

A similar situation exists in your body. Every cell in your body contains a membrane-bound nucleus which acts like a supercomputer and controls every action that occurs within a cell. Every cell in our body contains all the information required to produce an exact duplicate of ourselves. The program within the nucleus which contains all this information is encoded in the DNA.

Prokaryotic cells such as *E. coli*, whilst less complex than eukaryotic cells, also contain a cell wall. These prokaryotic cells also have many receptors (antennae) on the surface of the cell by which they sense the external environment. They have pores through which nutrients are imported and waste products are excreted. These cells also contain DNA which contains all the information required for the cell to function properly and duplicate itself. A major difference between eukaryotic and prokaryotic cells is that prokaryotic cells do not contain discrete compartments such as mitochondria. Also, the nucleus of the prokaryotic cell is not bounded by a membrane. Rather, the DNA floats around within the cell among the other cellular contents.

If the DNA contains the architect's plans for our cellular house, the question arises as to how the cell and the appliances within get built. In other words, what actually carries out the instructions encoded by the DNA? The computer nucleus may contain all the information, but it is obviously not possible for a computer, for instance, to get up onto the roof of a house and install a new TV antenna. You might imagine in our house that the computer instructs workers like maintenance men and women to install the appliances, paint the walls, fix the plumbing, remove the garbage, do the cleaning, etc. Alternatively, you might think that the computer controls a swarm of robots to do its bidding. Whatever the case, the point is that there must be 'workers' in biological cells which do the bidding of their DNA master. There are, and these 'workers' are called enzymes, which are made of a chemical called protein. Enzymes perform all the maintenance duties in the cell such as converting sugars into the energy-rich compound ATP. Many different enzymes are involved in the construction of new cells including the manufacture of all the cellular appliances. Numerous enzymes are also responsible for the maintenance of the DNA, to ensure that any fault in the super-computer nucleus hardware or software is promptly fixed.

One point that may need clarification with respect to enzymes: for the purposes of this book, you can regard all enzymes as being composed of protein. However, not all proteins are enzymes. Some proteins, for example, are regarded as structural (such as those which compose your hair) and whilst biologically important, do not carry out chemical conversions as is the case with enzymes. Some proteins are hormones—molecules with potent biological activity that can have profound effects on the functioning of an organism. Nevertheless, like enzymes, both structural proteins and hormones composed of proteins are encoded by DNA.

The point we are trying to make with all this is the following: DNA codes for enzymes, which are made of protein; enzymes are responsible for carrying out cellular function; all organisms are made of cells; and, thus, all organisms rely on enzymes for their function.

The building blocks that make up proteins are called amino acids, of which there are twenty. Proteins can contain as few as ten amino acids and as many as 5000. It is the number and arrangement of the amino acids in each specific protein that gives the protein the characteristics that allow it to undertake its specific function.

At this point, you can probably already see the link between all this and genetic engineering: genetic engineering relies on putting DNA from one organism into another or altering the DNA of an organism. Altering DNA alters the enzymes or proteins that the organism has, and can change biological function. Believe it or not, that is the biological basis of all genetic engineering. What makes genetic engineering so powerful is that even minor changes in the DNA that encodes a particular protein can have powerful effects on the function of that protein. Moreover, changing the function of only one protein can have powerful effects on the organism. You'll see some examples of this later.

DNA codes for proteins

You now know that DNA contains all the information required for the construction, maintenance and functioning of all life on Earth, and that all this information really codes for proteins which act as the 'workers' of the cell. What we haven't done is to examine the *mechanism* for how the DNA code is translated to produce proteins. By understanding this mechanism, you are in a very good position to understand the fundamental aspects of genetic engineering. Whilst we examine the mechanism below, for your interest we are also including a little extra information about how genes are arranged and are then trying to relate some of this information to things you might already know about.

Let's start by thinking about a human cell. By now you are probably thinking, 'Human → cell → multicellular → eukaryote → nucleus'. Every human cell (with some rare exceptions) contains 46 linear chromosomes (pieces of DNA) in the nucleus. The 46 chromosomes are organised as 23 pairs. The chromosomes contain the genetic information which is organised into thousands of different 'genes'. A gene is a stretch of DNA which codes for a particular protein, whether it is a structural protein (a protein that makes up part of a structure of the cell, for example, the cell wall) or an enzyme. Thus, genes are stretches of DNA which contain the genetic information to produce a protein. A single gene will encode a single protein.

At this point, it is probably worth thinking about the number of genes required to produce and maintain the functioning of an organism. In the case of higher eukaryotes, like humans, the total gene number has been estimated at about 100 000. Of the total number of genes, it has been estimated that about 10 000 of these genes are expressed (produce protein) in all cells. If you like, you could think of these genes, and their encoded proteins, as performing the housekeeping functions of all cells. We find this amazing: it only takes around 10 000 genes to direct the functioning of something as complex as a eukaryotic cell. Cells in humans perform different functions of course, so how many additional genes might it take to make a liver cell different from a brain cell? The answer: only about 1000–2000! So for each cell in your body, around 12 000 genes are expressed, around 10 per cent of the total number of genes present in each cell.

The structure of DNA

Whilst it may not be apparent to you at the moment, a study of how DNA codes for protein really has to start with a study of the structure of DNA. In fact, a study of the structure of DNA is also necessary to understand how this remarkable molecule replicates. But that is another story.

The structure of DNA takes the form of a double helix and is referred to as 'double-stranded DNA'. This structure could be thought of as a ladder which has been twisted at either end to form a helix. The sides of the ladder consist of repeating units of a sugar (deoxyribose) and a phosphate molecule, commonly referred to as the 'phosphoribose backbone' of DNA. If you don't understand the 'chemicalspeak' of the previous sentence, don't panic! The point is that the DNA molecule consists of two parallel backbones, analogous to the sides of a ladder.

The steps of the DNA ladder are made up of the four nucleotide bases: guanine (G), cytosine (C), adenine (A) and thymidine (T). Each base is joined to a deoxy-ribose sugar (backbone) and extends half-way across the step. This base is then paired to another base which extends from the opposite phosphoribose backbone. The nucleotide base A is always joined to (or paired with) T, whilst G is always paired with C (Figure 1.3). A convenient way to express the size of a DNA molecule is to talk about the number of 'steps' or base pairs that it has. There are 6000 million base pairs in nearly all of the five trillion cells that make up your body. It is the order of these four bases (G, A, T and C) which actually determines the order of amino acids in a protein and therefore the function of that protein. The code also includes information on when particular proteins and enzymes should be produced. This is an important point—imagine the computer in our analogous house telling the robot to take the garbage out on the wrong night! You may ask how the order of only four nucleotide bases can encode such complexity. The answer is simple when you consider other types of code systems you would be familiar with—Morse code, for example, and also remember that computers use a binary code, that is, everything in a computer is coded using only two digits, usually denoted by a '0' and a '1'.

The genetic code

We still really haven't finished yet, because we haven't told you how the genetic code is arranged. Each amino acid is encoded by a sequence of three bases, known as a 'triplet codon'. For example, the codon for the amino acid methionine is ATG. ATG just happens to be the start of the code for almost all proteins, so that almost all proteins, at least initially, begin with methionine. If the

Figure 1.3 The structure of DNA

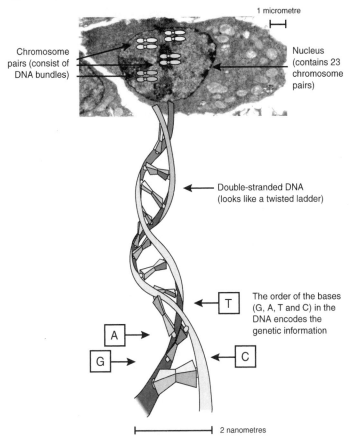

1 micrometre

Chromosome pairs (consist of DNA bundles)

Nucleus (contains 23 chromosome pairs)

Double-stranded DNA (looks like a twisted ladder)

The order of the bases (G, A, T and C) in the DNA encodes the genetic information

2 nanometres

next codon was GCT then it would code for the amino acid alanine. In this process, the protein's amino acid sequence would thus be methionine-alanine. The codes for all twenty amino acids have been worked out and are presented in Table 1.1. You can see from it that most amino acids have more than one code. This is because

Table 1.1 The genetic code

Amino acid[a]	DNA triplet codon[b]
Alanine (A)	GCT, GCC, GCA, GCG
Arginine (R)	CGT, CGC, CGA, CGG, AGA, AGG
Asparagine (N)	AAT, AAC
Aspartic acid (D)	GAT, GAC
Cysteine (C)	TGT, TGC
Glutamic acid (E)	GAA, GAG
Glutamine (Q)	CAA, CAG
Glycine (G)	GGT, GGC, GGA, GGG
Histidine (H)	CAT, CAC
Isoleucine (I)	ATT, ATC, ATA
Leucine (L)	TTA, TTG, CTT, CTC, CTA, CTG
Lysine (K)	AAA, AAG
Methionine (M)	ATG
Phenylalanine (F)	TTT, TTC
Proline (P)	CCT, CCC, CCA, CCG
Serine (S)	TCT, TCC, TCA, TCG, AGT, AGC
Threonine (T)	ACT, ACC, ACA, ACG
Tryptophan (W)	TGG
Tyrosine (Y)	TAT, TAC
Valine (V)	GTT, GTC, GTA, GTG
Stop codon[c]	TAA, TAG, TGA

Notes: a The one-letter abbreviation for each amino acid is given in parenthesis.
  b The triplet codons are always given based on the non-coding strand of DNA.
  c Stop codons signal the end of the protein.

there are 64 possible combinations of G, A, T and C when arranged in groups of three (the triplet codon).

Ribonucleic acid: the messenger of the cell

Whilst the impression we have given you that DNA codes for proteins by the arrangement of triplet codons is true, it is not the whole story. In fact, for any cell it is quite complicated to transform the DNA code into a functional protein. Rather than coding for proteins directly, the DNA message which codes for proteins is deciphered in two stages, called 'transcription' and 'translation'. DNA is 'transcribed' into a molecule called messenger ribonucleic

Figure 1.4 The DNA triplet codons are transcribed into mRNA and translated into protein

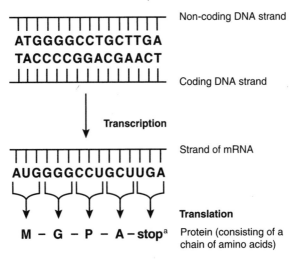

Non-coding DNA strand

Coding DNA strand

Transcription

Strand of mRNA

Translation

Protein (consisting of a chain of amino acids)

Note: a One-letter abbreviations for amino acids are given in Table 1.1.

acid (mRNA), which is related chemically to DNA, and then the mRNA is 'translated' into protein (Figure 1.4). mRNA molecules also have a backbone which is chemically similar to DNA, but the molecules almost always occur as a single-stranded form (that is, not in a double-stranded helix like DNA). As is the case with DNA, joined to the backbone of the mRNA molecule are four nucleotide bases, three of which—guanine (G), cytosine (C) and adenine (A)—are all found in DNA. However, whereas DNA has the fourth base thymidine (T), mRNA has uracil (U) as the fourth base. All this sounds rather complicated, but the point is this: the DNA code, which consists of combinations of G, A, T and C, is actually transcribed into a related code consisting of G, A, U and C, and it is this code which is used by the cell to finally produce the protein. (Table 1.1 shows the DNA triplet codons.)

Just to show you how wonderfully logical biology can be, we'd thought we would explain why this deciphering process is in two stages. Firstly, the machinery for producing proteins is not present in the nucleus, but outside the nucleus of eukaryotic cells. The mRNA is transcribed in the nucleus and then transported out of the nucleus to where protein synthesis occurs. DNA, on the other hand, never leaves the nucleus. The second reason for this two-step process has something to do with the cell being able to control production of specific proteins. Both prokaryotic and eukaryotic cells are able to switch genes on and off resulting in the presence and absence of the protein which the gene encodes. Specifically, the gene is either switched on and mRNA is being transcribed or it is switched off and the mRNA is not transcribed. Because the mRNA is slowly broken down by the cell, when a gene is switched off the mRNA is no longer produced and the mRNA which was previously produced when the gene was switched on is broken down, so that no more of that protein is produced. On the other hand, the DNA of the cell is never broken down because this would result in the loss of the genetic information. If DNA gave rise directly to protein without the mRNA intermediate, the cell would not be able to control protein production, because it cannot afford to break down the DNA blueprint. Let us now look in a little more detail at how the processes of transcription and translation occur.

Transcription of RNA

The process of transcription is undertaken by numerous enzymes that have evolved specifically for this task. Whilst we will not mention all the enzymes involved in transcription, we will describe the role of the most important enzyme in this process, which is called RNA polymerase. We have described DNA as a twisted ladder whereby the steps of the ladder are made up of base pairs (G paired with C and A paired with T) with one base of the pair extending from each side or strand of the DNA molecule. During the process of transcription, the DNA strands separate at the start of a gene. Imagine that each side of the ladder is pulled away from the other such that the DNA strands separate down the centre where the nucleotide base pairs are joined. This process is analogous to a zipper being unzipped. Then, the RNA polymerase sits on the start of the gene and starts to transcribe it, that is, synthesising an RNA molecule which is based on the DNA code of the gene.

Let's use a very simple example and assume that the DNA encoding the gene is only fifteen nucleotides long and consists of the DNA sequence TACCCCGGACGAACT (remember most genes are normally hundreds or thousands of nucleotides long). The RNA polymerase enzyme will allow an RNA nucleotide base to pair with the first nucleotide of the gene. In this case the RNA base A pairs with the gene base T. The next DNA base is A. RNA polymerase will allow the RNA base U to pair with the DNA base A. It will also join the RNA base U with the previous RNA base which was in this case the RNA base A. This process continues until the entire gene is transcribed into mRNA; in our example the RNA polymerase will synthesise the mRNA molecule AUGGGGCCUGCUUGA (Figure 1.5). In eukaryotic cells, this mRNA molecule will be transported outside the nucleus where translation takes place. In prokaryotic cells, translation will take place immediately.

Translation of mRNA into protein

The mRNA molecule is translated into protein by a very complex system involving a molecule called a ribosome. In fact, a ribosome is made up of many different proteins

Figure 1.5 Transcription of the imaginary fifteen-nucleotide gene

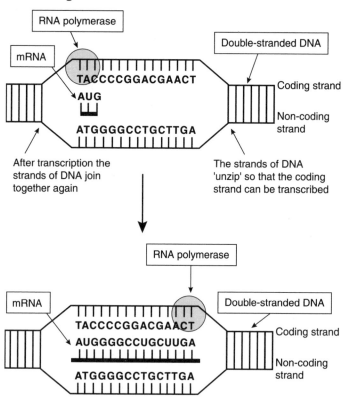

and also small pieces of RNA called ribosomal RNA. It is the ribosome which has the capacity to decode the nucleotide sequence and translate it into a protein (amino acid) sequence. The genetic code is translated such that three consecutive nucleotides (called a triplet codon) code for a single amino acid. Our imaginary fifteen-nucleotide mRNA sequence which was transcribed from DNA contains five codons: AUG-GGG-CCU-GCU-UGA. The first nucleotide codon AUG is translated into the amino acid methionine. The second codon codes for glycine, the third codon codes for proline, the fourth codon codes for alanine and the fifth codon codes for a 'stop signal' which signifies the end of the gene. There are 64 possible combinations in which the four nucleotide bases can be arranged. The amino acid or stop signal which each of these 64 codons encodes was given in Table 1.1.

We can now briefly describe the process by which the imaginary four amino acid sequence methionine-glycine-proline-alanine was produced by the ribosome. Firstly, the ribosome sits on the start of the mRNA molecule. Then the ribosome is able to identify the first codon of the gene. This first codon (called a start codon) usually codes for methionine. In the next step in this process, a transfer RNA (tRNA) binds to the start codon. A tRNA molecule is a short strand of RNA which can be bound to an amino acid. This process, by which the tRNA is joined to an amino acid, is also catalysed by specific enzymes. There are 61 specific tRNA molecules, each of which corresponds to a specific codon. Each of these 61 tRNA molecules is joined to the amino acid that is encoded by the specific codon. The tRNA which is bound to methionine contains the sequence UAC. This sequence of nucleotides is called the anti-codon. The ribosome will allow only the methionine-tRNA which contains the UAC anti-codon to bind to the AUG start codon in our imaginary mRNA molecule. Once again you will notice that the U base pairs with A and the G base pairs with C. The ribosome will then allow a second tRNA molecule (in the case of our imaginary gene, this is a glycine-tRNA molecule containing the anti-codon CCC) to bind to the second codon of the mRNA sequence (GGG in our gene). The ribosome will then join the methionine and glycine amino acids together. The ribosome continues to move down the mRNA sequence in

Figure 1.6 Translation of the imaginary fifteen-nucleotide gene

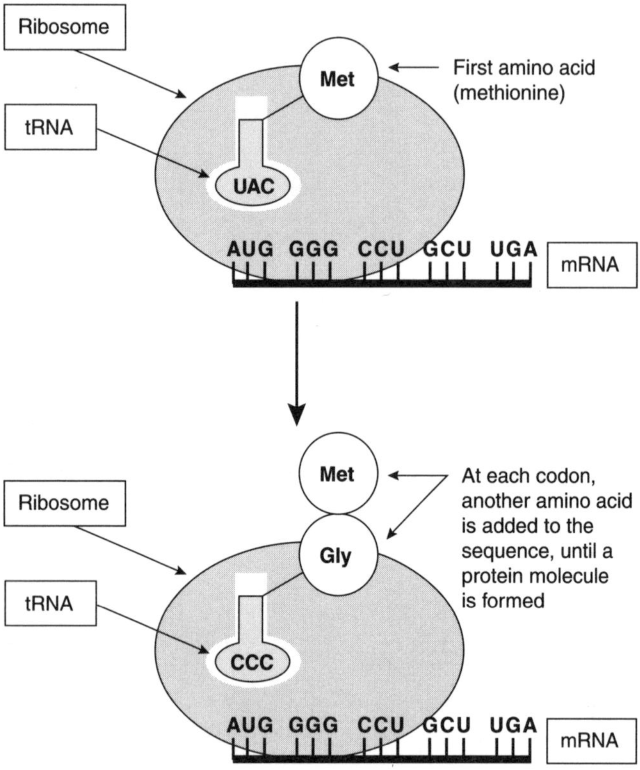

this way, translating the mRNA code into a protein sequence. When a stop codon is encountered, the ribosome falls off the mRNA molecule and releases the newly-synthesised protein (Figure 1.6).

After all this, your head is probably feeling like an overripe melon ready to explode! Even though you may not have understood all the above information, it doesn't matter too much as long as you have picked up the most important points. These points are:

- Genes encode proteins which are the 'workers' of the cell.
- Proteins consist of a linear sequence of amino acids.
- The organisation and control of proteins determine the function of a cell and ultimately the function of complex multicellular organisms like humans.
- Genes are a linear code contained on a molecule called DNA.
- DNA consists of two parallel strands twisted in a helix.
- Each strand has a backbone with a series of bases connected to it.
- It is this series of bases (G, A, T and C) which ultimately encodes proteins.
- The code is arranged in groups of three (a triplet codon). For example, ATG codes for the amino acid methionine.
- DNA is transcribed to mRNA.
- The mRNA, which is slowly degraded by the cell so as to allow the cell to control the expression of different genes, is translated into protein.
- Every protein of the cell, including RNA polymerase and ribosomes, is the result of transcription and translation.

Most importantly, the basic principle of genetic engineering is to alter the nature or amount of protein (usually an enzyme) in an organism by altering its genetic material. The logic of this process should be clear:

- Proteins are the molecules responsible for carrying out biological function.
- Proteins are encoded by DNA.
- Altering the DNA code alters protein function or the amount of protein.
- Altering protein function (or its amount) alters biological function.

At this point, you probably think that everything is a little surreal. We've talked about things like DNA, RNA and proteins like most people talk about cars, bicycles and going to the pub! How do you get an intuitive feeling for things like proteins and DNA and RNA when they are chemicals? You can't see the individual molecules, you can't touch them or smell them! The answer is not a simple one. Generally, we get a feel for these things by working with them and observing their properties under certain conditions. The following activities may help you better understand what protein and DNA are.

Activity one: isolating DNA

Ingredients and equipment:

½ cup wheat germ

salt

lemon juice

washing-up liquid

alcohol (a neat spirit from the drinks cabinet will do)

a tea strainer

You don't have to squint down a microscope to catch sight of DNA. Although animal DNA is tiny, you can easily extract DNA from plant cells in your own kitchen, in less than an hour.

First, mix about half a wine glass full of wheat germ with about 150 millilitres of cold water, in which you have dissolved ⅓ of a teaspoon of salt and two squirts of lemon juice. This should be gently stirred for about 10 minutes to break down the plant's cell walls. Press the mixture gently through the tea strainer and keep the pulp. Then repeat once more with the rest of the wheat germ.

Next, prepare 150 millilitres of cold water containing ⅓ of a teaspoon of salt, 3 teaspoons of alcohol and a couple of large drops of washing-up liquid. Add the pulp and stir. The detergent is now dissolving the DNA in the mixture. After 20 minutes of gentle stirring, add three heaped teaspoons of salt and stir the mixture for a further 10 minutes. Then leave it to stand until the solid settles out. Pour off the liquid and keep it. Throw the solid away.

Finally, dilute the liquid extract with about three times its volume of alcohol (clear spirits work best). As you stir, the plant DNA precipitates out as fine white threads which can be left to settle.

(Courtesy *New Scientist.*)

Congratulations, you have now isolated DNA! Principles similar to those used in this activity are involved in all spheres of genetic engineering. The next question is, 'What now?' What do you do with the DNA after you isolate it? The answer to this question is discussed in the next chapter.

Activity two: thinking about the properties of proteins

- Proteins consist of amino acids linked together. There are twenty different kinds of amino acids which make up proteins, and they all have different chemical properties. There are some you actually come across in everyday life. Why don't you look around the shops for them? Here are some amino acids (or their combinations) you may find: monosodium glutamate, aspartame, L-lysine. Why don't you taste them? Dissolve them in water and taste them again. What colour is the solution? You can imagine that inside a cell, these amino acids can also be floating around waiting to be incorporated into proteins.
- Gelatine is a commercial preparation used in making jellies. It is actually prepared by boiling tendons and skin. These tissues contain a protein called collagen.

Collagen is a structural protein, not an enzyme, and consists of long thread-like molecules. When these molecules are pulled apart (denatured) by boiling, rather than reforming when they cool, they link up with water molecules in a tangled array called a gel. Why don't you make some jelly and think about the processes involved?

Egg white consists of a large concentration of a globular protein called albumin. Albumin is structurally related to other globular proteins like the enzymes. You can use egg white as a model to think about the properties of enzymes. If you whip egg whites they form a creamy substance which, when combined with sugar and baked, forms pavlova. What is happening to the molecules of albumin? As you whip them, the globular albumin unfolds (denatures) and forms a tangled array with itself. Incorporated in this tangled array is air and water from the egg white—thus a foam is formed. Adding sugar to the foam makes it taste nice. The baking process usually consists of putting the mixture into a moderate oven, the heat from which causes the air in the foam to expand, making the pavlova rise. The oven is then usually turned down to a lower temperature which doesn't really 'cook' the pavlova, but dries the outside, forming a crust.

What happens when you boil an egg? Can you see that the albumin can unfold (due to heating rather than whipping, as is the case with pavlova) and form a matrix? No air this time though, so something akin to a gel is formed.

Activity three: observing the properties of enzymes

Enzymes carry out a bewildering array of chemical reactions. You can see how an enzyme works using easily available materials.

You will need:

a small piece of cabbage

peroxide solution (hydrogen peroxide, which you can get from a pharmacy, is used for bleaching hair)

Procedure:
Before you add the cabbage, look carefully at the hydrogen peroxide solution. You should notice that small bubbles are produced. These bubbles are the result of the hydrogen peroxide breaking down to form water and oxygen. Oxygen is a gas and forms the bubbles. The breakdown of hydrogen peroxide to water and oxygen is a chemical reaction. All chemical reactions will speed up if they are heated—you may notice that the amount of bubbling increases as the solution warms, and explains why you should keep the solution in the fridge.

Now add the cabbage. You should see a vigorous bubbling. The reaction has sped up considerably. This is because cabbage contains an enzyme called catalase, which increases the rate at which hydrogen peroxide is broken down. This experiment tells you a lot about enzymes—they function to increase the rate of chemical reactions.

By the way, enzymes are being increasingly used commercially. An example is the use of the enzyme lipase. This enzyme breaks down fat and is used in washing powder. The enzyme is actually produced by genetic engineering: the gene encoding the lipase was isolated and manipulated to produce large amounts of the enzyme.

Conclusion

That's it for our look into the basic principles of biology that you need to grasp in order to understand genetic engineering. By now you've seen that the basic principle of genetic engineering is to add or alter genes to change proteins to change the function in organisms. The first question you should probably ask then is, 'How do you get your genes in the first place?' As Mrs Beeton once wrote at the beginning of her recipe for rabbit pie: 'First catch your rabbit'. As you can see from the recipe above for isolating DNA, the principle is not difficult to understand—cells are ground and broken, separated, and the DNA is precipitated from solution using alcohol (or some other chemical). We'll examine this process in more detail in the next chapter, but then go a few steps further: after you have isolated your DNA, what do you do with it? Like Mrs Beeton's rabbit, the DNA is also cut up, but there the similarity ends.

2 Cloning a gene: first catch your DNA, then cut it up . . .

If you remember back to the beginning of Chapter 1, we gave you a pretty rough-and-ready definition of genetic engineering: taking a piece of DNA (which contains a gene) from one organism and putting it into another organism. You have now seen that the actual process of obtaining DNA from an organism is, in principle, not a difficult one—it involves simply grinding tissue of the organism such that cells are broken open, and then subjecting the cellular contents to a number of chemical and physical treatments. In other words, the first part of our definition, 'taking a piece of DNA', seems relatively simple. In fact, DNA was first isolated by the Swiss scientist Frederick Miescher in 1869.

So if isolating DNA from organisms is not a problem, why did it take until the latter part of the twentieth century to develop genetic engineering? What we hope you will see in this chapter is that although isolating DNA is not normally difficult, the 'putting DNA into another organism' and, more importantly, *keeping* it in the recipient organism can be pretty tricky. Moreover, the actual detection of a fragment of DNA which contains *the specific gene* you want can be quite difficult. By looking at how DNA fragments are isolated, put into recipient organisms and maintained in them, we will be able to illustrate the basic principles behind genetic engineering, help explain how genetic engineering works and also explain why it took so long after Miescher first discovered DNA for genetic engineering to become possible.

To start our examination of the process of putting DNA into other organisms, let's just briefly review what this actually means in terms of what you learnt in Chapter 1. Firstly, you'll recall that DNA is a complex chemical compound which consists of a backbone (sugar phosphate) to which are attached a number of bases. These bases are of four kinds, designated by the letters G, A, T and C, and it is the sequence of letters arranged in units called genes which specify proteins—molecules that are responsible for ultimately performing biological function. Thus the essence of genetic engineering lies in the ability to identify genes, located on fragments of DNA, and then transfer specific fragments of DNA that encode particular proteins to another organism. Let's look at one example of how this works.

The role of genetic engineering in the treatment of diabetes

The example we have chosen involves the bacterium *E. coli* and its use in the production of one protein, insulin, for the treatment of diabetes. Before we start explaining how the fragment of DNA containing the insulin gene was found and put into *E. coli*, we should explain a little about diabetes.

Diabetes is a relatively common disease which varies in severity and affects about 2–3 per cent of the population. The name 'diabetes' means 'siphon' and was used by the Greeks over 2000 years ago because, in severe cases of the disease, despite huge food intake marked weight loss occurs—the body's substance was thus believed to be dissolving and pouring out through the urinary tract (a belief which was not far from the truth). One of the salient features of diabetes is the prodigious output of very sweet urine, hence the full name of the disease, diabetes mellitus ('mellitus' meaning 'sweet'). Diabetes mellitus can be a very serious disease indeed: in England in the seventeenth century, the disease was colourfully referred to as the 'pissing evil'. The reference to evil is not an understatement as untreated diabetes can produce very serious symptoms such as marked weight loss, blindness, arteriosclerosis (hardening of the arteries, which leads to heart attacks and poor circulation, which may eventually require limb amputation), brain damage, coma and death.

From the above description, you may not find it surprising that people with diabetes suffer from a number of problems with their blood chemistry, one of the most significant being elevated levels of the sugar glucose in their blood. What is the cause of diabetes mellitus? There are a number of possible causes, but one of the most frequent involves the body producing insufficient quantities of a protein called insulin. Insulin, which belongs to a class of molecules called hormones, is responsible for producing a number of effects in the body, one of which is the stimulation of the body's cells to take up glucose. Thus, a lack of insulin, which is one of the major causes of diabetes, results in a failure of the body's cells to take up glucose, which results in elevated levels of blood glucose. Luckily, diabetes mellitus is treatable; diabetics are able to monitor the glucose levels in their blood regularly, and can control their blood glucose levels by regulating their diet and injecting themselves with insulin, thereby avoiding the harsher symptoms described above and allowing 'insulin-dependent' or 'type I' diabetes sufferers to lead a relatively normal life.

Where does the insulin used by diabetics to control this disease come from? Well, insulin is a protein which is present in minute quantities in blood. Theoretically it would be possible to extract insulin from human blood, but this process yields only small quantities of insulin which would not be sufficient to fulfil the daily requirements of diabetics. You would also be aware of the possibility of contamination of insulin purified from human blood with viruses which cause diseases such as AIDS and hepatitis. Because of these problems, in the past insulin was isolated from pigs. The use of pig insulin in the treatment of diabetes was not without its problems. Pig insulin is slightly different to human insulin, that is, the sequences of amino acids which make up the two proteins are somewhat different. This difference meant that diabetic patients' immune systems could sometimes recognise pig insulin as foreign to their bodies and build an immunity to it, much like the rejection of transplanted organs in transplant patients. Along similar lines, pig insulin may produce allergic reactions in diabetics, which can be very severe. Also, prolonged use of pig insulin often leads to a decrease in its effectiveness; this problem is also caused by the pig insulin being slightly different to its human counterpart. Hopefully by now you can see how genetic engineering could be useful. Wouldn't it be great if you could take the human insulin gene, put it in another organism, like the bacterium *E. coli* (which is easy to grow and doesn't harbour viruses that could affect humans), and grow large quantities of the insulin-producing bacterium and then harvest the insulin? In fact, this is what's done. The human insulin gene has been taken from one organism (in this case humans) and put into another organism (the bacterium *E. coli*), which then produces the human insulin protein, which is easily extracted and purified for use by diabetics. The use of human insulin produced by genetic engineering clearly has many advantages: the insulin is human, not pig, and therefore does not produce allergic reactions and other problems. Moreover, by producing human insulin in *E. coli*, which can be grown in big vats like those used for brewing beer, vast quantities of insulin-producing bacterial cells can be produced, from which insulin can be easily extracted. Of course, there is no requirement for the use of human blood, thus avoiding the possibility of contaminating human insulin with blood-borne viruses.

So how was the human insulin gene isolated and put into *E. coli*? You already know something about the first part of the answer: human DNA can be extracted by chemical and physical procedures. But what about the next steps? After you have extracted human DNA, how do you put the DNA into *E. coli*? You are also aware that human DNA contains about 100 000 genes: how do you find the DNA fragment that contains the human insulin gene? Keep in mind these critical questions as we guide you through the five steps to clone a gene which we have outlined below and summarised in Figure 2.1.

Step 1. Catch your organism and isolate its DNA

Firstly, catch your organism, be it human, mouse, plant or bacterium. It is obvious how you would catch a mouse—you simply need to be able to run fast enough. To catch a plant, you don't even need to be able to run very fast, just sneak up on it slowly. For a human, it's probably a good idea to ask permission first before you take a blood sample to prepare DNA. You know the rest; grind up the tissue and extract the DNA using a procedure similar to that outlined in Chapter 1. In fact, you don't even need very much material to prepare lots of DNA, a few leaves from a plant or less than 1 gram of

Figure 2.1 The five steps to clone a gene

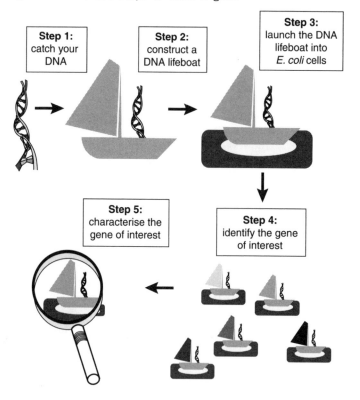

mouse tissue or a thimbleful of human blood will do. In fact, it is technically possible to obtain enough DNA from a single human hair.

At this point it is probably worthwhile spending some time explaining how you catch a bacterium. You will remember from our description of *E. coli* in the previous chapter that bacteria are single-cell organisms that cannot be seen with the naked eye. How can you catch something that is too small to see? Well, the answer to this question lies in the fact that, whilst one organism is difficult to see and manipulate, a large number growing together (as a colony) can be easily manipulated and isolated. One can see this to some extent with respect to everyday food spoilage. The scum you see on the top of milk, or the growth you see on solid materials like cheese or bread, represent colonies of bacteria and mould made up of millions of single-cell organisms. Obviously, the growth of contaminating microbes on foodstuffs is slowed by keeping these materials refrigerated. This is not surprising since, as a rule-of-thumb, bacteria grow fastest at higher temperatures, and many grow best between 20°C and 37°C.

The above discussion actually identifies those principles that are used to isolate, grow and manipulate bacteria; the media used to grow bacteria are essentially foods containing things like sugars and other nutrients such as vitamins. The medium can be a liquid (as milk was in our example above), which is called a broth, or a solid material (as cheese was in our example). This solid material is normally broth which has been solidified with a gelling agent called agar. This same material was, and still is, used as a gelling or solidifying agent in cooking, particularly in Asian cuisine. After adding the bacteria, the broth or agar medium is generally left at 20–37°C for between one and seven days for the bacteria to grow (Figure 2.2).

Isolating DNA—or is that RNA?

As we've previously illustrated, isolating DNA is in principle very easy. However, when scientists are trying to clone a gene from eukaryotic cells, they often do not isolate DNA from the organism but instead isolate messenger RNA (mRNA). (Recall that DNA serves as the template to produce mRNA which then serves as the template to produce protein.) After mRNA is extracted from cells, it is then converted to DNA (in this case, the

Figure 2.2 Uninoculated and inoculated cultures of *E. coli* grown in broth and on agar plates at 37°C for 24 hours

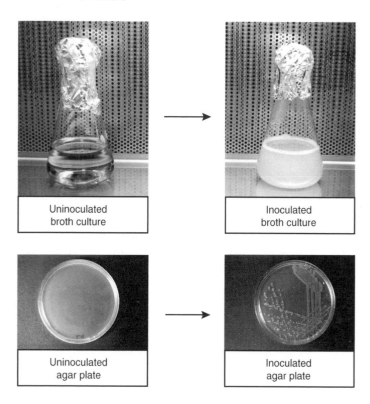

DNA is called complementary DNA or cDNA). This process is carried out using an enzyme called reverse transcriptase.

You're probably thinking that such an approach is crazy. Why go to all the trouble of isolating mRNA (which isn't as easy as isolating DNA) and then turn the RNA into DNA? Why not isolate the DNA directly? There are two reasons for using this strategy rather than simply isolating the DNA directly from cells. Firstly, you will remember from the previous chapter that each human cell, for example, contains about 100 000 genes, of which only 10 000 genes in any given tissue are transcribed into mRNA and translated into protein. If you know that the gene you wish to clone is expressed (or produced) in, for example, liver cells, then the chances of finding the gene improve from 1 in 100 000 to 1 in 10 000 by extracting liver cell mRNA and converting it to cDNA. If the gene is expressed at high levels, then there will be lots of copies of the mRNA which will further improve the odds of finding the correct gene. Secondly, unlike prokaryotic genes, eukaryotic genes usually contain introns, which are sequences of nonsense code that interrupt the protein-encoding regions of the gene. Thus, the RNA encoded by a gene containing introns includes both information for the protein sequence and pieces of gobbledegook. Does this matter? Yes! Ribosomes, whose job it is to translate the mRNA sequence into protein, are stupid and don't 'know' that there are introns in the mRNA and will quite happily translate the mRNA sequence, gobbledegook and all.

Because of this intron problem, eukaryotic cells have to splice the introns out of RNA before the mRNA can be translated. Prokaryotes do not have this problem and thus do not have the machinery to carry out this splicing function. The bottom line is that if you want to put a eukaryotic gene in a prokaryotic cell (like putting the human insulin gene into *E. coli*) and get a sensibly functioning protein, you must first cut out the introns. You don't do this of course: you let the eukaryotic cell do it for you (splice out the introns of its mRNA) and then you isolate the mRNA and turn it into DNA. In other words, the cloning of cDNA gets around this 'intron problem' (Figure 2.3).

Figure 2.3 Introns are removed from eukaryotic genes during the production of cDNA

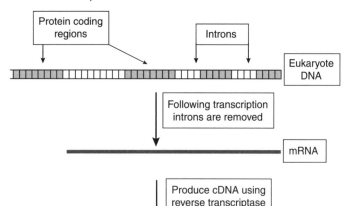

Step 2. Production of a recombinant DNA molecule: 'DNA lifeboats'

So now you can imagine that you have extracted DNA from an organism, or extracted mRNA and made cDNA. If you recall, we were using the cloning of the insulin gene as an example. So of course, in this example, because humans are eukaryotes, you would have extracted mRNA and made cDNA. You're probably thinking that the next logical step would be to find the fragment of cDNA which contains the insulin gene and put it into *E. coli*. Not so! For technical reasons, which will become apparent to you as we progress through these steps, it's actually easier to put all the cDNA fragments into *E. coli* cells and then look for the fragment you want.

'All right,' you say, 'let's put the DNA into the *E. coli* cells.' Not so fast! Life wasn't meant to be easy. Before you put DNA into any organism, there is another problem

Figure 2.4 The process of binary fission in prokaryotic
(*E. coli*) cells

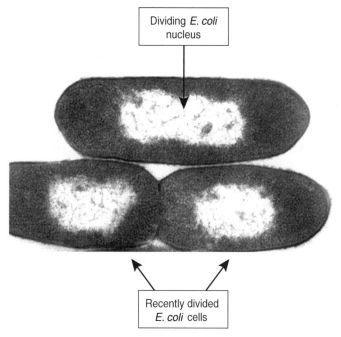

Dividing *E. coli*
nucleus

Recently divided
E. coli cells

you must get around. It's not so obvious, but we need
to think about it. We'll call this the dividing-cell problem.
To explain what this problem is, we'll use *E. coli* as an
example.

The dividing-cell problem

E. coli, like many other bacteria, replicates by a process
called binary fission. All this really means is that one cell
splits evenly into two daughter cells, they both grow in
size, and then the two daughter cells split into two, giving
four cells and so on (Figure 2.4). You may well ask what
this has to do with gene cloning. Well, the problem with

cell division is that cells must replicate their DNA before they split into two. In the case of *E. coli*, as with most other bacteria, their genetic material is mostly comprised of one large circular DNA molecule called a chromosome which is about three million base pairs in size. If the chromosome is not replicated during cell division, after one division each daughter cell would only contain one half of the original genetic material (1.5 million base pairs), after two cell divisions, one quarter (750 000 base pairs) and so on until the DNA is diluted out of existence. Can you see the potential problem with genetic engineering? Simply 'putting DNA into *E. coli*' will not work; the DNA which you introduce must replicate along with the dividing cells or it will be diluted out of existence.

You're now probably asking, 'Why not just put DNA into non-dividing cells of *E. coli*?' The answer is that you can, but that brings with it even more problems. For instance, it is normally only possible to put DNA into a handful of cells, thus such an approach would not allow you to grow vast quantities of cells in huge vats to produce a protein like insulin. Also, non-dividing cells tend not to be so metabolically active and thus do not produce large amounts of protein products anyway.

At this stage we hope you can see that this problem with dividing cells is a big one. In fact, the same problem applies to all cells, both prokaryotic and eukaryotic. Maybe you've already thought of a solution: introduce the foreign DNA into the chromosome of the *E. coli* cell. That way, as the cells divide, the chromosome of the cell replicates, and the foreign DNA can be piggybacked along with it. This solution is, in fact, a good one and is often used in gene cloning, particularly with eukaryotic cells. However, for technical reasons too involved to go into, this technique is often not the most satisfactory.

There is another solution to the dividing-cell problem; this solution still involves the idea of using *E. coli* DNA (in this case, though, the *E. coli* chromosome is not used) to piggyback foreign DNA and was actually used in the first gene cloning experiments in 1973. In this year, a paper carrying the rather sober title 'Construction of biologically functional bacterial plasmids *in vitro*' and authored by Stanley Cohen and colleagues was published in the journal *Proceedings of the National Academy of Sciences (USA)*. This paper actually heralded the start of what we now call genetic engineering. What was in this modestly titled paper that had such an impact? Well, it's all got to do with DNA molecules called plasmids. It is plasmids that were used to piggyback foreign DNA and solve the dividing-cell problem.

What are plasmids? Plasmids are small, circular fragments of DNA, which replicate inside bacteria independently of the main chromosome. There are many different types of plasmids, some of which contain genes encoding physiological functions, antibiotic production, antibiotic resistance, virulence (ability to cause disease), bacteriocins (chemicals that kill other bacteria) and toxin production. Some plasmids even have genes that encode their ability to transfer themselves to other bacteria (this ability partly explains why bacteria often quickly become resistant to antibiotics). However, in contrast to bacterial chromosomes, which contain most of the bacterial genetic material (around three million base pairs), plasmids are very small, some of them containing only around 3000 base pairs. In one respect though, plasmids are no different to bacterial chromosomes: because of the problem of cell division, plasmids, like the main chromosome, have to replicate to survive—no replication means dilution out of existence (Figure 2.5).

What Cohen and his colleagues reported was that they were able to take plasmids out of *E. coli* cells and join them with foreign DNA in the test tube (they'd made recombinant DNA) and place them back into *E. coli*. The

Figure 2.5 Plasmids replicate in unison with cell division

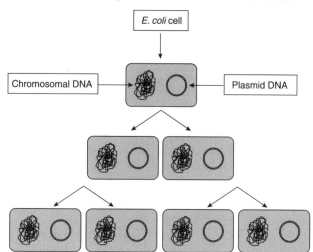

plasmids then replicated along with the dividing cells, piggybacking the foreign DNA. The dividing-cell problem was solved!

We've come a long way from 1973, and there are now a multitude of different plasmids that are used for different purposes in gene cloning projects. You shouldn't lose sight of the fact, however, that these plasmids have been constructed by humans for one principal task: to piggyback foreign DNA and provide a solution to the dividing-cell problem. To distinguish these artificial plasmids from their naturally occurring counterparts, we often refer to them as plasmid vectors or simply vectors. Moreover, there are also other 'vectors' (DNA molecules used to clone genes) which aren't plasmids. However, for the sake of simplicity, in this book we intend to talk only about plasmid vectors.

One simple way to think of plasmids is to regard them as a lifeboat for genes when introduced into the host cell:

Figure 2.6 Some features of plasmids: an origin of replication sequence, an antibiotic resistance gene and restriction enzyme sites

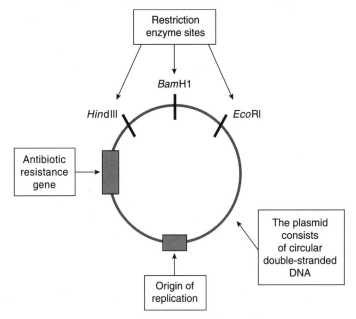

if the DNA you wish to clone is not safely joined to the plasmid–DNA lifeboat, it will 'sink without trace' when introduced into the host cell (you know of course that this is because of binary cell division). At this point, given the obvious importance of plasmids in gene cloning, it's worthwhile talking a little more about some of their features (Figure 2.6).

Firstly, you already know that plasmids are able to replicate inside cells. This is because plasmids contain a sequence of bases called an 'origin of replication' which allows the plasmid to be replicated with the dividing host cell. You should also understand that cells can afford to harbour a lot of plasmids because they are relatively small molecules—it costs energy to replicate DNA, and the smaller the DNA molecule, the less energy it takes to make and retain copies of it. Thus, cells containing hundreds of plasmids per cell can be grown. This means that when you extract plasmids from cells, you can get a very high yield of plasmids, which makes working with them easier. Moreover, because plasmids can exist in cells in multiple copies, the foreign DNA, which is spliced into the plasmid, exists in cells as multiple copies as well. This means that there is more protein produced—more copies of a particular gene actually means more messenger RNA which means more protein (which is what you want).

Secondly, plasmids contain antibiotic resistance genes. Antibiotic resistance genes code for protein products that protect the bacterium from being killed by the corresponding antibiotic. For instance, the gene that codes for resistance to penicillin produces an enzyme called beta-lactamase, which is able to break down penicillin. The presence of antibiotic resistance genes in plasmids allows you to select for the presence of the plasmid once it has been transferred into recipient cells. One simply has to add penicillin to the growth medium to ensure that only those cells which harbour a plasmid (that carries a penicillin resistance gene as well as the gene you want to isolate, for example, the insulin gene) are able to grow. (You probably don't understand the full significance of antibiotics yet, but things will become clear when you read the section below on transformation.)

Thirdly, plasmids contain 'restriction enzyme recognition sites'. These are the sites in the plasmid which are recognised and cut by restriction enzymes (described below) and into which the genes that you are cloning are inserted.

At this point you're probably wondering how plasmid DNA and foreign DNA are cut and spliced together. How do you cut up DNA? Actually, this question is really,

'How do you cut DNA into very tiny fragments?' Remember these fragments have to be small enough to contain a gene or a few genes which exist as submicroscopic entities. The answer is that you don't use scissors or very sharp razor blades, but enzymes that act like molecular scissors. There are many types of enzymes that do this, and they are referred to collectively as 'restriction enzymes'. Their special feature is that they recognise specific sequences of bases in DNA and, where those sequences exist, they cut the DNA molecule just like a pair of molecular scissors. Each restriction enzyme recognises and cuts only one specific base sequence—its own restriction enzyme recognition site. For instance, the restriction enzyme *Pst*I recognises the sequence CTGCAG, the restriction enzyme *Bam*H1 recognises the sequence GGATCC and *Sau*3A recognises the sequence GATC. So you can imagine what happens after you have extracted your DNA; you add a restriction enzyme to it and the enzyme cuts that DNA into small fragments.

How small are they? Well, it's actually pretty easy to figure out. Take, for example, *Pst*I. This enzyme recognises a sequence of six bases: CTGCAG. Since sequences this small are distributed more or less evenly amongst DNA, you can do a very easy statistical calculation: the chance of the first base being C is one in four (the other possible bases being T, G or A). The chance of the second base being a T is also one in four, and of course the calculation proceeds: $4 \times 4 \times 4 \times 4 \times 4 \times 4 = 4096$. In other words, on average, the chance of any given sequence of six bases occurring is one in 4096. Thus *Pst*I will cut DNA about once in every 4096 bases.

So now you can imagine that you have the DNA you want to clone cut up into small pieces. Of course, you now know that plasmid DNA is a circular molecule and that you have to splice DNA into it. This means you must cut your plasmid DNA as well. Normally, you would

Figure 2.7 The formation of recombinant DNA

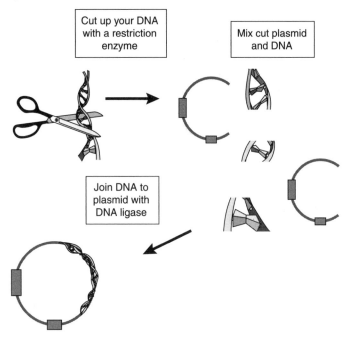

cut your plasmid using the same restriction enzyme you used to cut up the DNA containing the gene you wish to clone. Plasmid vectors containing only one site for the restriction enzyme you are using are generally chosen.

So now we have plasmid DNA and foreign DNA which has been cut with a restriction enzyme. The next step is to join the molecules together. Again, we need an enzyme to do this. The molecular welding machine used for this purpose is an enzyme called DNA ligase, which is able to ligate (join) pieces of purified DNA and plasmid vector together (Figure 2.7). In fact, at the risk of confusing you, it's worthwhile explaining that ligase will blindly join together any loose ends of DNA. For example, ligase will join back together a plasmid which has been cut by a restriction enzyme, forming the original uncut plasmid, or, for that matter, any two DNA fragments. So you can see that after adding ligase to a tube containing many types of DNA fragments, many types of combinations will result producing quite a heterogeneous mixture.

To summarise, to produce a recombinant DNA molecule containing a human gene, you would take purified human DNA (or human cDNA) and a plasmid vector, add some of the selected restriction enzyme to cut them both up, then ligate the chopped up DNA and plasmid vector together using DNA ligase. Once this process is completed, the ligation mixture will contain many molecules of plasmid vector which have ligated back together by themselves, plasmid vector which contains foreign DNA inserts, and even human DNA fragments joined together. An important point to remember is that only a small proportion of the plasmids will contain the gene you wish to clone because all the other genes that are coded by the DNA will also have been independently cloned into the vector plasmid. It is now time to transfer this heterogeneous mixture of recombinant DNA molecules into the recipient cells.

Step 3. Transfer of recombinant DNA molecules into recipient cells: launching the DNA lifeboats

There are several methods by which recombinant DNA molecules can be transferred into *E. coli*. However, the most popular method used by scientists is called chemical transformation. The process of chemical transformation involves growing *E. coli* cells in broth, then collecting the cells and mixing in a calcium chloride (a type of salt) solution, which is then left on ice overnight. This treatment allows the *E. coli* to become 'chemically competent': the cells are now able to take up DNA, which is added to the *E. coli* calcium chloride solution. After adding recombinant DNA molecules and leaving this mixture on ice for one hour, the cells are suspended in nutrient broth culture and warmed at 37°C for three hours. This three-hour incubation allows plenty of time for any cells which have taken up plasmid DNA to express the antibiotic genes carried by the plasmid.

By now you're probably wondering what sort of tangled mess the gene cloning method has got itself into: you produce this great mixture of joined DNA fragments, only some of which are plasmids that contain the gene of interest; then you transform *E. coli* with this mixture, and only some of the cells will be transformed with DNA. Don't worry, this is where you start untangling this mess to find the *E. coli* cell that harbours a plasmid which contains the gene you are looking for.

The first step is to sort out the cells that have been transformed with plasmid DNA from those that haven't. This step is pretty simple—you add your transformation mixture onto agar medium containing an antibiotic. You will remember that plasmid vectors normally contain a gene or genes which code for antibiotic resistance. Thus only cells which contain either the original plasmid vector or a plasmid vector with DNA inserted into it can grow in the medium that contains antibiotics, since only these cells are antibiotic resistant.

So how do you then work out which colony is the one containing the DNA of interest? There are a number of ways to do this, and this usually represents the most difficult part of the gene cloning process. Rather than explain how you would find the insulin gene, which is a pretty complicated process, we're going to use another more simple example first. We also thought it might be useful to describe this example starting with the initial steps in the cloning procedure (namely extraction of DNA) to recap all the ideas we have covered so far.

Figure 2.8 Results of an experiment involving the cloning of the violacein genes from *Chromobacterium violaceum* into *E. coli* using a penicillin resistant plasmid vector

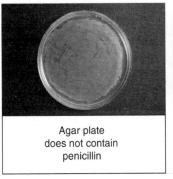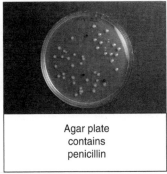

Agar plate does not contain penicillin	Agar plate contains penicillin

The purple dye violacein is produced by the bacterium *Chromobacterium violaceum*. The DNA from *Chromobacterium violaceum* has been isolated, digested with restriction enzymes and ligated into a plasmid vector which contains a penicillin resistance gene. (In this case you wouldn't make cDNA because *Chromobacterium* is a prokaryote and doesn't contain introns.) The recombinant DNA molecules were then transformed into chemically competent *E. coli* and spread over the surface of agar plates. As you can see in Figure 2.8, when the transformation mixture is placed on agar which does *not* contain penicillin, the plate is covered with bacterial growth which is a mixture of mostly *E. coli* cells that don't contain plasmids, some *E. coli* cells that contain the vector plasmid only, and a very few *E. coli* cells that harbour plasmids which contain the violacein genes. However, when the same *E. coli* cells that have been transformed with the recombinant DNA molecules are placed on agar which *does* contain penicillin, then those *E. coli* cells that do not contain a plasmid are unable to grow. Each colony of *E. coli* cells which is able to grow on penicillin medium contains the plasmid vector.

Because the *E. coli* have been placed on a medium containing penicillin, only those cells transformed with the plasmid (which contains the antibiotic resistance gene) are able to grow and produce colonies. Now, the question is, 'Which colonies contain the cells with the violacein genes?' The answer is, 'Those that are purple!' Those *E. coli* cells that harbour a plasmid containing the violacein genes produce enzymes which act to produce a purple colour (violacein) and are thus purple in colour. The bottom line is that the simplest way to pick which clone contains the recombinant plasmid you want is to merely look at it!

This above example illustrates several points about gene cloning. DNA (or cDNA) can be isolated, digested with restriction enzymes and ligated into a plasmid vector to form recombinant DNA molecules. The mixture of plasmids from a typical gene cloning experiment can be transformed into chemically competent *E. coli* cells and grown on agar plates containing antibiotics that select for only those cells which contain the plasmid vector. Finally, it is easy to clone genes for pigments (if the genes are properly expressed, meaning proteins coded for by the cloned genes are produced in *E. coli*) because you can see the difference in *E. coli* cells which contain the pigment genes and those that don't.

However, as with most things, gene cloning isn't that simple all the time. Frequently, the proteins expressed by cloned genes are not easily detectable. Sometimes, the genes are present but they don't produce any protein. So you sometimes end up with a plate containing many colonies, one of which may be the one you want, but it is visually indistinguishable from all the other colonies. This is normally the case, and generally the biggest hurdle in genetic engineering is to identify the *E. coli* cells harbouring the vector that contains the genes you wish to clone. In the example we started with, trying to clone the insulin gene, finding the cells that contain this gene is like trying to find a needle in a haystack.

Step 4. Identification of *E. coli* cells containing the gene of interest: the needle-in-the-haystack problem

Given the right facilities and some expert guidance, it might be possible for you to clone every human gene. This collection of human genes would be ligated into a plasmid vector and transformed into chemically competent *E. coli* cells. The problem is that, because there are 100 000 human genes, you will have amassed millions of separate *E. coli* cell colonies, each containing different pieces of human DNA. All of these *E. coli* colonies will look identical. So how do you identify the *E. coli* cells that contain the insulin gene, the cystic fibrosis gene or a gene which may potentially provide a cure for cancer from amongst all the other *E. coli* cells that you have constructed?

The answer to this question is very complex. For example, it took over five years of research by hundreds of scientists (say, the equivalent of 500 years of work for a single person) to identify the gene involved in cystic fibrosis. There are also many different strategies that can be used to identify a specific gene once it has been cloned into *E. coli*. For the sake of simplicity we will only describe one of the most important methods by which cloned genes can be identified—the development and use of an oligonucleotide probe (an oligonucleotide is a short single-stranded DNA sequence).

The first step in this process is to purify the protein product of the gene you wish to clone. Protein purification is in itself a tricky business. Generally, cells are broken open and different chemical and physical procedures are applied in order to extract the protein. Because the success of this approach relies on the chemical characteristics of the particular protein, and because all proteins are chemically different, there are hundreds of different methods by which scientists go about purifying proteins. We don't want to fill this book with information about protein purification methods, but we would like to highlight two very important points. Firstly, if you want to purify a specific protein, you need a method for confirming that your chosen protein is present in a heterogeneous mixture of other proteins. If you can detect the protein you are trying to purify, then you can follow the success or otherwise of your protein purification procedure. Such methods generally rely on being able to detect the specific enzymatic activity of the protein you are trying to isolate. Scientists have devised many different ways to detect many different proteins which, for example, can rely on the conversion of a colourless substrate to a coloured product by the enzyme you are trying to purify. Secondly, despite your best efforts it may be impossible to purify the protein or even develop a method by which to detect the presence of the protein you wish to purify. In this case you will have to consider using an alternative strategy to clone your selected gene.

However, let's assume that you have been successful in purifying your protein (if you're very lucky this task will have taken less than six months of your time). What do you do with it? Well, what you can do is put your purified protein into a machine called a protein sequencer. As the name suggests, this machine can determine the amino acid sequence of the first part of the protein—normally up to the first twenty amino acids. Let's say that the protein sequencing machine tells you that the first seven amino acids of your purified protein are methionine-cysteine-aspartic acid-glutamine-lysine-asparagine-isoleucine. It is now possible to design an oligonucleotide probe based on this amino acid sequence. If you turn back to Table 1.1, you

Table 2.1 Designing an oligonucleotide probe: working out the nucleotide sequence

methionine–cysteine–aspartic acid–glutamine–lysine–asparagine–isoleucine						
AUG	UGU	GAU	CAA	AAA	AAU	AUU
	or	or	or	or	or	or
	UGC	GAC	CAG	AAG	AAC	AUC
						or
						AUA

can figure out what nucleotide base sequence might code for this seven amino acid sequence (Table 2.1).

You can see from Table 2.1 that working out the DNA sequence that encoded the seven amino acids is not completely simple, because most amino acids have more than one triplet codon. Thus, there are 96 ($1 \times 2 \times 2 \times 2 \times 2 \times 2 \times 3$) different DNA sequence possibilities for these seven amino acids, of which only one will be absolutely correct. It is now possible to program another machine, called a DNA synthesiser, which is able to synthesise each of these 96 DNA sequence possibilities (or at least the most likely possibilities). Each of the newly synthesised oligonucleotides is then joined to a chemical 'label' so that the presence of the oligonucleotides can be detected. Usually the chemical label is a radioactive atom (phosphorus[32]) which can be detected because it will cause a black spot to form on X-ray film. Alternatively, the oligonucleotides can be joined to a chemical called biotin which, when mixed with a specific enzyme and substrate, will result in a coloured product which can be seen by the eye. Either way the oligonucleotide probes are now said to be 'labelled'.

Have you guessed yet how you can use your labelled oligonucleotide probes to detect the gene you wish to identify? Remember that this gene, which codes for the purified protein (including the seven amino acids at the start of the purified protein), has been joined to a plasmid which is sitting in a single *E. coli* colony. Remember also that this *E. coli* colony looks identical to a multitude of other *E. coli* colonies that you have generated. The answer lies in the fact that normally only the *E. coli* colony which harbours the gene you wish to clone will contain the DNA sequence that codes for those seven amino acids. Some of the *E. coli* cells from each and every *E. coli* colony you have generated can be lysed (split open) and the DNA exposed. By increasing the pH, the double-stranded DNA in the *E. coli* cell (including the plasmid DNA which may contain the cloned gene) will break apart and form single-stranded DNA. When the labelled oligonucleotide probes are added, some will be able to bind to the complementary DNA sequence of the gene that codes for the seven amino acid sequence. The oligonucleotide probes will not be able to bind to DNA which does not code for the seven amino acid sequence. The labelled oligonucleotide probes bound to the gene you wish to clone can then be detected by applying and then developing an X-ray film or by adding the specific enzyme and substrate that binds to biotin, which will result in a coloured product which can be seen by the eye (Figure 2.9).

If everything has gone really well with their experiments, most scientists would be very happy to have cloned a specific gene within twelve months. However, there are a lot of pitfalls in this process. And that is why, when a scientist working in this area has successfully cloned a gene (signified by a dark spot on a piece of X-ray film), it is not uncommon to see them jumping around the lab waving their arms wildly with a big grin on their faces.

Step 5. Characterisation of the gene of interest: what do you do with the gene you have just cloned?

Finally, you have cloned the gene. Following the celebrations and a brief recovery period, you will wish to

Figure 2.9 Use of an oligonucleotide probe to detect a gene of interest

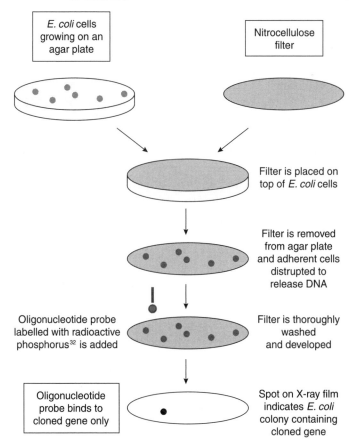

characterise the gene. 'Characterise' in this sense means finding out information about the gene such as its sequence of bases, what restriction enzymes can cut it, how much protein is produced from it and whether you can place the gene in other plasmids to get better expression. There are actually many things that you can do with your newly found gene, and we are not even going to try and list them all for you—a popular book that describes in detail methods used by scientists to clone and characterise genes is called *Molecular Cloning* and was written by J. Sambrook, E.F. Fritsch and J. Maniatis. This book comprises three volumes and weighs in at 3.7 kilograms!

There are, however, two very important things you would probably want to do to your newly cloned gene that you should be aware of. The main thing you would want to determine is the sequence of bases which makes up the gene. The way in which DNA sequencing is performed is way beyond the scope of this book, so we won't describe it here. However, having this sequence is very useful. For example, you can use the sequence to identify sites where restriction enzymes will cut and which can be used to transfer the gene to other plasmid vectors (see below). Once you know the DNA sequence of your cloned gene, it's also possible to optimise the expression of the gene so that it is highly expressed in *E. coli* or other prokaryotic or eukaryotic organisms. This is generally carried out by positioning signals (DNA sequences) for high level transcription and translation at the start of the gene. These expression signals are normally found within an 'expression plasmid vector'. The newly identified gene can be cloned into a restriction enzyme site next to the transcription and translation signals which are contained in the expression vector. The expression vector containing the cloned gene can then be transferred into the correct host organism.

How do you actually cut your gene from the plasmid vector, isolate it and combine it with other plasmids such as expression vectors? Firstly, you can extract the recombinant DNA plasmid from as little as 1 millilitre of *E. coli* broth culture, and produce millions of copies of the plasmid.

Next the recombinant DNA molecule is cut up with restriction enzymes. Normally, you know the DNA sequence of both the vector and the cloned DNA, so you can select restriction enzyme sites which flank the cloned DNA. By cutting with the correct restriction enzymes, you can cut the cloned DNA away from the vector DNA. If this is the case, you then have two fragments (vector and what was formerly the insert). The problem is now to separate them and get your hands on the insert.

This is possible using a technique called 'agarose gel electrophoresis'. In this process, the two types of fragments (vector and insert) are separated by employing an electric current to drag them through a gel which has very small pores. As these fragments are dragged through they separate from one another on the basis of their size. Thus, when DNA is introduced into one end of the agarose gel (into slots formed in the agarose gel) and electrophoresed (an electric current is applied), because DNA is negatively charged it will migrate towards the positive electrode. The smaller the DNA molecule, the faster it migrates. Thus, the mixture of DNA molecules will be separated on the basis of size.

The DNA fragments can then be stained using a specific dye called ethidium bromide so that they are seen as 'bands' in the gel (Figure 2.10). The DNA bands are made up of many copies of the same-sized DNA fragment. The next step is to simply cut out the part of the gel which contains the band of DNA that you want and extract the DNA fragment using well-established methods. You can now take your expression vector, cut it with the right restriction enzymes and ligate in your isolated fragment.

What is in the expression vector that makes it so special? Well, what sets expression vectors apart from other vectors is that they have signals (sequences of bases) for high levels of transcription and translation which you

Figure 2.10 Agarose gel electrophoresis and staining of plasmid DNA

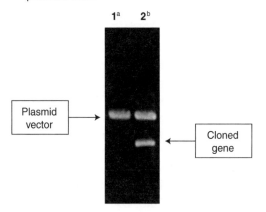

Notes: [a] Lane 1 = Plasmid DNA isolated from *E. coli* cells containing only the plasmid vector.

[b] Lane 2 = Plasmid DNA isolated from *E. coli* cells where a cloned gene has been inserted into the plasmid vector.

can place next to the start of your gene. What you do of course is choose the correct restriction enzymes that cut both the insert and the expression vector such that when they are joined together, the gene you are cloning is the correct distance away from these signals. The expression vector could be one that is active in either a prokaryotic or a eukaryotic host. Thus, by selecting the right vector you can clone a prokaryotic gene in a prokaryotic or eukaryotic host or a eukaryotic gene in a eukaryotic or prokaryotic host.

How do you know that the host organism is producing the protein you want? One way of finding this out is to use SDS-polyacrylamide gel electrophoresis. This technique works in a fashion similar to that of agarose gel electrophoresis but uses polyacrylamide as the gel rather than agarose. Using this technique, the proteins of the cell

Figure 2.11 SDS-PAGE separation and staining of proteins

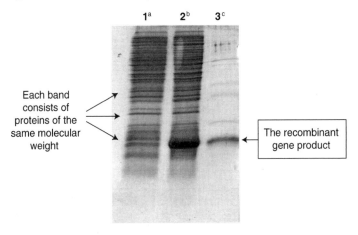

Each band consists of proteins of the same molecular weight

The recombinant gene product

Notes: ^a Lane 1 = Proteins produced from *E. coli* cells containing only the plasmid expression vector.
^b Lane 2 = Proteins produced from *E. coli* cells where a cloned gene has been inserted into the plasmid expression vector.
^c Lane 3 = Purified recombinant protein.

can be separated according to their molecular weight and stained a blue colour using a dye called Coomassie blue. Notice in Figure 2.11 that the *E. coli* cells which contain the expression plasmid that doesn't contain the cloned gene don't express the protein product of that gene. On the other hand, when the cloned gene is present in the expression plasmid, the protein product of the cloned gene is highly expressed and can even be purified. As was the case with agarose gel electrophoresis, the protein bands that can be seen in Figure 2.11 represent many individual proteins of the same size which are therefore found at the same position in the SDS-polyacrylamide gel.

Now your head is probably *really* feeling like an overripe melon ready to explode! Try not to let that happen because, firstly, it would result in a terrible mess, and secondly, we have now finished dealing with the most complex concepts of how you clone, identify and characterise a gene. The following chapters are the fun parts, detailing the many and varied uses of genetic engineering. Although you may not have understood all the above information, it doesn't matter too much as long as you have picked up the most important points. These are:

- DNA can be isolated from eukaryotic or prokaryotic cells (see Activity one in Chapter 1).
- Plasmid vector DNA can be similarly isolated.
- Restriction enzymes cut DNA whereas DNA ligase joins DNA fragments together to form recombinant DNA molecules.
- Recombinant DNA molecules need a way of replicating: they need to be inserted into plasmids or chromosomes.
- Recombinant DNA molecules are transferred into cells, as in the transformation of *E. coli*.
- *E. coli* containing the gene of interest can be identified using techniques such as the use of labelled oligonucleotide probes.
- Genes can be expressed and the protein product purified.

Conclusion

By isolating plasmids and introducing recombinant DNA molecules into host cells, Cohen and his colleagues had started a revolution. They had shown it was now theoretically possible to take *any* piece of DNA—from another bacterium, plant, mouse, monkey or human—and splice it into plasmid DNA and replicate it in bacteria. In other words, the technology was available to transfer DNA across previously impenetrable species barriers. The shackles were broken, no longer were we confined to shuffling genes within species by breeding. Like the revolution in flight caused by the Wright brothers, breaking the species barrier revolutionised gene technology. We had a new concept in biology—transgenic or recombinant organisms, that is, organisms which contain foreign DNA. The penultimate sentence in Cohen and his colleagues' groundbreaking paper is a classic of understatement:

> The general procedure described here is potentially useful for insertion of specific sequences from prokaryotic or eukaryotic chromosomes or extrachromosomal DNA into independently replicating bacterial plasmids.

By now you will have gained a fairly good understanding of the methods used by scientists to clone, identify and characterise a gene. Hopefully you are now starting to get an inkling of why gene cloning is so revolutionary—you can take genes from one organism and introduce them into another. This technology will in some way touch all our lives in the not-too-distant future. In fact, when you read some of the examples of the uses of genetic engineering we shall give in the next chapters, you may find that this technology has already benefited you or members of your family.

3 Production of recombinant protein using single cells

At the end of the last chapter, we said that you might discover that genetic engineering has already affected your life or the lives of your family without you realising it. This is partly because it is generally not immediately obvious that a product is the result of genetic engineering. Most of the time we don't even consider how a product has been manufactured (particularly with pharmaceutical products, where people often don't know exactly what the drugs they are taking are or how they work, let alone what process is used to produce them), and the actual process of manufacture is not usually included on the label information. For these reasons, and because there has been some controversy and anxiety over the use of gene technology (as discussed in Chapter 1), those responsible for marketing genetically engineered products usually don't mention the process of manufacture when advertising or labelling the products. This is a pity, because some of the most remarkable accomplishments of the technology are left hidden in the fine print, which in turn can lead consumers to think that this information is being hidden for more sinister reasons.

What are some of the accomplishments of gene technology? Well, here is a list of the things that prokaryotic and eukaryotic cells can be used to produce: pharmaceuticals, restriction enzymes, amino acids and vitamins, antibiotics, polymers, enzymes, antibodies, vaccines, insecticides and nitrogen compounds (by nitrogen fixation from atmospheric nitrogen). Alternatively, microbes in particular can be used to break things down. The most obvious examples of this are their use in the breakdown of toxic chemicals (for example, oil spills), and the breakdown of sugars and cellulose to produce things like alcohol and some organic acids.

There are many recombinant products produced by prokaryotic and eukaryotic cells, but let's limit our discussion by narrowing the field and restricting our examination to the human protein pharmaceuticals. In the past, as in the case of insulin, for example, the production of such proteins has been very difficult, and normally only limited quantities of proteins were available for therapeutic use. The production of protein pharmaceuticals by genetic engineering has now improved the range, efficacy and availability far beyond what one could ever imagine prior to the development of such technology. Indeed, the world market for protein drugs has been growing at an annual rate of 12 to 14 per cent. Of course, we talked about the production of human insulin in the last chapter. Are there any other examples? Well, cancer patients undergoing chemotherapy are treated with a human protein called granulocyte macrophage colony stimulating factor, or GM-CSF for short. This protein, which is produced by genetic engineering, stimulates the body's defence systems to quickly recover after being weakened by chemotherapy so that patients are less susceptible to infectious diseases. Another example is the use of protein hormones, derived

Table 3.1 Recombinant proteins used for treating human disease

Product	Use
Erythropoietin	Anaemia and kidney disorders
Insulin	Diabetes
Human growth hormone	Growth hormone deficiency
Factor IX	Haemophilia
Tissue plasminogen activator	Thrombolytic agent
Relaxin	Childbirth facilitator

from genetic engineering, to promote conception in women in *in vitro* fertilisation programs. Listed in Table 3.1 are some other products derived from genetic engineering and their uses. Some of these products have already been approved by the US Food and Drug Administration for use in humans. So, clearly, there have already been major accomplishments with the use of genetically engineered products with obvious potential for growth in the future.

So, if you've already cloned a gene encoding a useful and important protein pharmaceutical and want to produce large quantities of it for the world market, how is this done? Clearly, the problem is no longer one of genetic engineering—the gene is already cloned in a vector and is present in an organism like *E. coli*. The answer, obviously, is to grow large quantities of the organism in question and then purify the protein of interest. From what we discussed in Chapter 1, you should recall how prokaryotes like *E. coli* or, for that matter, simple eukaryotes like yeast can be grown—either by inoculating liquid growth medium (broth) with the organism, or plating them on the same or similar medium which has been solidified with agar. In this chapter we'll give you an idea of how such systems, in particular those employing broth, which has the potential to produce the greatest number of cells, can be scaled-up to produce large quantities of recombinant cells and their protein products of interest. We'll then examine some concrete examples of these processes.

Large-scale fermentation

Before we examine how we can produce recombinant proteins on a large scale, let's look at what is meant by fermentation. Fermentation can be described as a process whereby chemical changes can be brought about by the action of prokaryotic or eukaryotic cells. The word comes from the Latin word 'fermentare' meaning 'to cause to rise', and originally referred to the processes of beer and wine production and the leavening of bread—all of which employ yeast.

Let's start our examination of large-scale recombinant protein production by imagining that you have recently produced in the laboratory a recombinant *E. coli* strain containing a plasmid that includes our gene of interest. Such a recombinant system would obviously be initially produced in what we could term laboratory-scale quantities—that is, the quantities used to easily grow the organism and produce sufficient recombinant protein so that you could measure and assess the efficacy of the protein and its subsequent production scale-up.

What would these quantities be? The volumes of growth media involved would range from a few millilitres to a few litres. Growing the organism would thus involve the dispensation of media into flasks, inoculation with the organism and incubation at 37°C (the optimal growth temperature of *E. coli*). In addition, the inoculated medium would normally be shaken, as this aerates the medium and supplies oxygen to the cells, allowing them to grow quickly and produce a greater number of cells for a given volume of medium. This last point is not a trivial one—aeration of the culture medium and an adequate supply of oxygen to growing cells is essential in obtaining high cell densities and yields of recombinant proteins—the more cells you have, the more recombinant protein you can produce.

Given the above points, one could think, understandably

but somewhat naively, that to scale-up (to produce more organism and hence more protein), you merely need to expand what already takes place in the laboratory. Unfortunately this is not the case. For a start, it would not be cost-effective to set up 20 000 individual 1-litre flasks to obtain 20 000 litres of cell suspension. Alternatively, you could set up one big vat of 20 000 litres. On the face of it, this approach seems quite feasible—beer is brewed in such vats all the time.

Unfortunately, this simple solution will not work because of the difficulties in aerating large volumes of cell suspension—a pre-condition for producing large concentrations of cells. Sufficient aeration is not required for the brewing of beer and is, in fact, undesirable: beer is brewed without oxygen (anaerobically), since this condition forces the yeast to convert sugar to alcohol and carbon dioxide. However, anaerobic conditions do not allow for efficient growth because they inhibit efficient utilisation of fuel—a condition contrary to that needed when producing a recombinant protein.

So here is the problem, in a nutshell, of using large volumes of material and achieving sufficient aeration: the rate at which the system is aerated is dependent on the ratio of the surface area of the cell suspension (as this is the only way through which oxygen can get to the medium), the volume of the medium and the rate of shaking. A rough rule of thumb in the laboratory is that aerobic conditions exist when the flask is around ten times the volume of the medium employed (high surface area to volume ratio) and is shaken at 250 rpm. So you can see that to scale-up such a system to 20 000 litres is almost impossible: you would need a 200 000-litre fermentation vat (200 cubic metres—about the size of a large lounge room!) and be able to shake it at 250 rpm.

So if you can't simply scale-up from the same conditions employed in the laboratory, what can you do?

Figure 3.1 A typical fermentation system used for the production of recombinant proteins

ARC Biotech's 1400 square metre pilot plant houses 34 highly instrumented fermenters ranging in scale from 20 to 15 000 litres. (Photo courtesy of Alberta Research Council, Canada.)

The solution is simple—you bubble either air or oxygen through the cell suspension. Although there are many different examples of such systems, called bioreactors, they all rely on this simple principle. Figure 3.1 shows one type of bioreactor.

You can now see the bottom line for how large-scale production of a recombinant protein can be achieved. However, you should also keep in mind that insufficient aeration is not the only thing that can limit cell concentration. Other things, such as the depletion of nutrients as the culture grows, will also serve to limit cell concentration. This problem can be solved to some extent by the continual addition of nutrients in what is called a fed-batch process.

As you can see, there are many and varied aspects to the employment of large-scale fermentation processes and, not surprisingly, whole books can also be found on this subject.

Using a genetically engineered protein to treat heart attack patients

Have you had a heart attack lately? We certainly hope not, but the odds are that as you grow older you or someone you know may suffer a heart attack (one in every five deaths in Australia is caused by heart disease). What is a heart attack? Well, a heart attack is caused by a blockage in the arteries of the heart. Usually the blockage is caused either by a build-up of cholesterol and other fatty substances (called an atherosclerotic plaque) or by a blood clot (thrombosis). Eventually, this blockage stops the blood supply to the heart which results in the intense pain experienced during a heart attack and damage and death of the local heart muscle. When brought to a hospital, heart attack patients are treated with a human protein called tissue plasminogen activator (TPA). TPA acts as a blood thinning agent to break down blood clots (which may also form as a result of the initial heart artery blockage) and stop such clots from provoking further heart attacks. TPA is a product of genetic engineering—so how was the TPA gene cloned and expressed to produce the large quantities needed to treat heart attack patients?

The cloning of the TPA gene was accomplished by the use of an oligonucleotide probe. First human TPA was purified and the amino acid sequence at the start of the TPA gene determined. Then an oligonucleotide probe based on this amino acid sequence was synthesised and labelled. This probe was found to bind to the human TPA gene, which had been ligated into a cloning vector and introduced into *E. coli*. Once the genes were cloned, the DNA sequence of the TPA gene was determined.

Figure 3.2 The treatment of heart attacks using tissue plasminogen activator (TPA)

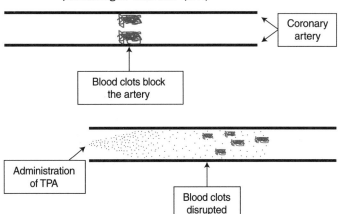

The TPA gene was then transferred into expression vectors. In fact, TPA can be expressed either in *E. coli* or in eukaryotic cells. These cells can be grown in a fermentation vat and large amounts of TPA can be purified for use in heart attack patients as a blood thinning agent (Figure 3.2).

Producing a safer whooping cough vaccine

Whooping cough is a disease of the respiratory tract which is particularly severe in the young. In fact, if you have the opportunity to talk to someone who worked in the medical profession prior to the 1950s, ask them about whooping cough. You will probably be told that hospital wards were filled with infants suffering from this disease, who coughed so much that they became short of breath and hence had to rapidly intake air with a 'whooping sound', giving the disease its name.

Whooping cough could, and did, result in many childhood deaths. The cause of this disease is the bacterium called *Bordetella pertussis*. By the time it is diagnosed, the bacterium has generally already released the toxins that result in the disease, making antibiotic treatment less than effective. However, an effective treatment against this horrible disease was developed in the 1950s. This treatment was called a 'whole cell vaccine'.

What is a whole cell vaccine? Well, basically, *Bordetella pertussis* cells are grown in a big fermentation vat as described above, then the bacterial cells are killed by chemical treatment. Once the *Bordetella pertussis* cells have been killed, they are formulated into a vaccine and used to immunise children. When this vaccine has been injected, the cells of the body which are responsible for protecting us from disease get much needed practice in eliminating the dead *Bordetella pertussis* cells. The trick in this process of vaccination is one of inducing memory in the cells of the immune system: once these cells have recognised *Bordetella pertussis* once, the immune system 'remembers' that they are foreign and subsequent attacks by this microbe are met with much stronger defence. This type of process—of inoculation with a killed or incapacitated pathogen and the production of immunological memory—forms the basis of many vaccination procedures.

Now whilst vaccinations have provided one of the greatest revolutions in medicine and public health, there exist situations where, for varied reasons, vaccinations are not effective or are dangerous—and vaccination against *Bordetella pertussis* is no exception. In this case, the effectiveness of whooping cough vaccination is dependent on every child in the population being vaccinated. This is the case because, firstly, infants are not protected from infection by *Bordetella pertussis* until after the vaccination regimen is completed (normally not until six months of age). This means there is a pool of very young infants in any population that are not protected against the disease. Secondly, some infants that have been vaccinated with the whole cell vaccine are *not* protected from infection by live *Bordetella pertussis*. If the entire population has been vaccinated, then these 'at risk' infants are not really at risk, because there are so few of the disease-causing bacterium circulating in the population. If, on the other hand, the vaccination rate in a population starts to drop, then the *Bordetella pertussis* bacterium starts to circulate in the population and outbreaks of whooping cough start to occur, primarily in the unvaccinated and 'at risk' infants.

Unfortunately, the vaccination rate against whooping cough has been dropping since the mid-1970s, particularly in some developed countries. This is because the whole cell whooping cough vaccine is associated with side effects. Mild side effects (which are also the most common) include redness and swelling at the site of injection. Moderate side effects (which occur less often) include persistent crying. Finally, experts in the field still argue as to whether the whole cell whooping cough vaccine is responsible for severe side effects such as brain damage or death (which may only occur one time in a million). Experts do agree that the benefits of vaccination with the whole cell vaccine clearly outweigh the risks of side effects. Despite this expert opinion, vaccination rates have dropped and, not surprisingly, the incidence of whooping cough has increased.

So how can genetic engineering help with this problem? Well, scientists have discovered that only three *Bordetella pertussis* proteins are required in a vaccine to protect infants from catching whooping cough. These three proteins were purified and formulated into a vaccine called an 'acellular vaccine' which protects children from catching whooping cough and also decreases the number of reported side effects by 80 per cent. One of the three proteins is called pertussis toxin, a very complex protein made up of five subunits which is responsible for several of the symptoms of whooping cough. The other two

Figure 3.3 Pertussis toxin is converted into pertussis toxoid by changing two amino acids

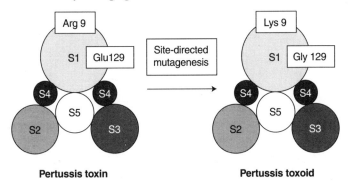

Pertussis toxin → **Pertussis toxoid**

proteins are called pertactin and filamentous haemagglutinin. Obviously it is not possible to incorporate a toxin into a vaccine for infants. However, it is important for the cells of the immune system to practise against this toxin, so that when the infant is exposed to *Bordetella pertussis*, the immune system can eliminate both the *Bordetella pertussis* and the pertussis toxin.

How was this dichotomy resolved? Well, the pertussis toxin genes were firstly cloned. This was accomplished (once again) by the use of an oligonucleotide probe. Yes, you guessed it, the pertussis toxin was initially purified and the amino acid sequence at the start of the S1 subunit of pertussis toxin determined (pertussis toxin consists of five different proteins called S1–S5 in a ratio of 1:1:1:2:1). Then an oligonucleotide probe was synthesised and labelled, and this was found to bind to the pertussis toxin genes which had been ligated into a cloning vector and introduced into *E. coli*. Once the genes were cloned, the DNA sequence of the pertussis toxin genes was determined. Next, the DNA sequence of the pertussis toxin genes was changed so that two very important amino acids were replaced in the S1 subunit sequence (Figure 3.3).

These two amino acid changes in the S1 subunit result in the toxin being 'knocked out' so that it is no longer toxic (such knocked-out toxins are called toxoids). The method used to make the changes in the DNA sequence is called 'site-directed mutagenesis'. We won't describe here how site-directed mutagenesis is done as it is very complex and is beyond the scope of this book. The knocked-out genes, which now code for pertussis toxoid, were introduced into *Bordetella pertussis*, replacing the original pertussis toxin genes. The method used to replace these genes is also very complex so we won't try to describe it here either. Nonetheless, the result of all this genetic tinkering was a *Bordetella pertussis* bacterium which produces the non-toxic version of pertussis toxin (pertussis toxoid), as well as pertactin and filamentous haemagglutinin.

This genetically engineered *Bordetella pertussis* bacterium is now being grown by vaccine companies in fermentation vats to produce lots of pertussis toxoid, pertactin and filamentous haemagglutinin. These proteins are purified and incorporated into the new acellular whooping cough vaccine, which protects children from catching whooping cough and also decreases the side effects by 80 per cent.

Cleaning up the environment: PCB degradation

In the examples of genetic engineering given so far, the common theme has been that some particular product, for example a protein such as TPA or insulin, has been produced. However, genetic engineering can do much more than just produce a range of different protein products for specific uses. If you cast your mind back to Chapter 1, you will recall that genes can encode enzymes and that these enzymes carry out a multitude of chemical transformations in cells—to produce energy, break down particular molecules and construct particular molecules. The ability of enzymes to break down molecules is of particular interest in this section. We can illustrate just how effective genetically modified organisms can be in this regard using the example of PCB degradation.

PCBs (polychlorinated biphenyls) are toxic compounds which were produced up until the 1970s for a diverse range of uses. For example, PCBs were used in transformer fluid, capacitors, carbon paper and even lipstick. When it was realised that PCBs were in fact toxic, their production stopped, but this was not the end of the problem. For a number of reasons, not the least of which being that these compounds did not exist in nature prior to their production by humans, these compounds are very persistent in the environment. Thus, although the production of PCBs has been halted, they still present a problem. This problem of persistence contrasts markedly with naturally occurring substances, which are generally broken down fairly quickly by the action of microorganisms when released into the environment.

To explain how genetic engineering may help in the problem of PCB contamination, we must first briefly examine what 'broken down in the environment' means. As discussed above, most compounds in the environment are broken down by microorganisms. 'Breaking down' means chemically converting a particular compound into another. In the case of toxic compounds like PCBs, this chemical conversion turns the compounds into less toxic forms. As you may have guessed, this conversion process, carried out by microorganisms, is really the result of the action of an enzyme or, more usually, the action of a number of enzymes acting in a concerted fashion, changing the compound in a series of steps to a less toxic form.

Now it just so happens that some microbes in the right circumstances can break down PCBs, but there are some problems with this process. One of the main problems is that the genes encoding the enzymes which break down PCBs are not switched on in the presence of PCBs—obviously a major problem.

So how can genetic engineering help? If you could clone the genes encoding the enzymes responsible for the degradation of PCBs, you could then place these genes under the control of special genetic switches which allow you to manipulate exactly when you want the genes to be 'turned on'. Moreover, you could put such genetic units into any microbe you want, for example, one that is particularly good at surviving in an aquatic environment or in soil. If one has the genes encoding PCB degradation from a number of microorganisms, some of which degrade different PCBs, you could even combine these genes in organisms to produce more efficient degraders.

Now whilst there is still work to be done on PCB degradation, this example does illustrate some important principles, the most important being that it is possible to take genes and combine them in certain ways, such that novel or more efficient metabolic pathways can be constructed. (A metabolic pathway is a series of chemical conversions carried out by enzymes.) This holds true not only for pathways that can break down materials, but also for those that build them up. For example, metabolic pathways have been altered to produce a number of vitamins and amino acids.

Conclusion

As you can see from the examples given in this chapter, both prokaryotic and eukaryotic cells can be used as factories for the production of recombinant proteins. These proteins, for example, enzymes, vaccines and hormones, are currently being used in many different fields of human endeavour, most notably in human medicine, where many lives are being saved through their use. Cells can also be genetically modified so that aspects of their metabolism (their ability to convert chemicals) are altered. Such alterations in metabolism can enhance the cells' ability to destroy or break down undesirable compounds, such as toxins, or enable those cells to produce important complex chemicals like some amino acids and vitamins.

In this chapter we have looked at the production of recombinant proteins in cells grown in fermentation vats. In the following two chapters we will investigate the use of whole multicellular organisms (transgenic plants and transgenic animals) for the production of recombinant proteins.

4 Transgenic plants: solar powered genetic engineering

In the previous chapter, we examined the use of unicellular organisms (bacteria, yeast or mammalian cells) as biological factories for the production of recombinant proteins. As we discussed previously, there are some clear advantages in employing organisms and cells of this type: producing genetically engineered single cells tends to be relatively simple (particularly in the case of prokaryotes and yeast) and, given the appropriate fermentation technology, these organisms are relatively easy to grow to produce large quantities of cellular material and recombinant protein.

Having said that, there are also some disadvantages in employing single cells as biological factories. It would not have escaped your notice, for example, that such technology often relies on the use of quite sophisticated fermentation and cell-culture techniques to produce cell concentrations of sufficient density. Indeed, from the discussion in Chapter 3 and the reference to 'whole books being published on fermentation', you would have gathered that we simplified our description of this area of scientific and technological work considerably. Therefore, considerable attention has been given to seeking other ways to cheaply and more easily produce recombinant proteins or other products. One solution has been to employ plants as biological factories. At first, this idea might sound somewhat peculiar, but a moment's reflection should reveal that it makes very sound sense because it is relatively easy to propagate and then harvest large fields of plants using relatively simple, yet well established, technology.

In this chapter we would like to explore just how plants can be employed as biological factories and explain how genetically engineered plants are produced, giving you some examples of what has been accomplished in this field.

What is a transgenic plant?

A transgenic plant is one which is genetically modified and as such is a multicellular organism which contains a foreign gene in each of its cells, in much the same way that a culture of a bacterium, such as *E. coli*, can harbour a cloned gene in each of the many cells that make up that culture. As we have previously established, it is the expression of this foreign gene that results in the production of protein and which mediates biological function.

There are two main reasons why scientists have developed methods for producing transgenic plants. Firstly, if you can clone genes into plants, you may be able to modify particular traits in those plants. Not surprisingly, a major focus in plant genetic engineering has been on ways to improve agricultural productivity, but there have been other interesting developments—for example, the development of blue roses. Secondly, as already described above, plants can be used as bioreactors, that is, as a means to produce a particular protein or product without having to resort to complex fermentation technology. Let's now look at how transgenic plants are produced. Later we will explore in detail a few examples of transgenic plants and their uses.

The production of transgenic plants

As you may recall from Chapter 2, the process by which unicellular organisms such as *E. coli* are made to take up DNA is called transformation. The idea of transformation was to introduce foreign genes (which code for proteins/ enzymes that mediate biological function) into the *E. coli* cell. Because *E. coli* is unicellular, once you have transformed one cell and selected that cell using antibiotics, you can then grow vast quantities of these identical *E. coli* cells.

It seems almost intuitive that this approach to transformation is not going to work with plants because they are multicellular organisms. It is difficult to imagine how one could treat a whole plant in the same way as *E. coli* to make it take up DNA. Moreover, if you recall, the process of transformation was quite inefficient, meaning that even if you could subject a whole plant to this process, only a very limited number of cells would take up the foreign gene. And even if you produced a partly transformed plant, which did produce a foreign protein, there is still the greater problem of how this plant could be propagated. If you were, for example, going to propagate the plant by seed (which is convenient and one of the reasons why producing transgenic plants is so attractive in the first place), unless the transformed cells are germ cells (reproductive cells that will develop into seed), plant propagation methods involving fertilisation and the production of seeds would give rise to offspring which *do not* carry the foreign gene.

Thus, just because the organism you are trying to

transform is multicellular, additional problems arise that are not encountered when one is dealing with unicellular organisms. So the question is, 'How do you get around this problem of plant multicellularity and the introduction of foreign genes?' Well, the solution relies on the unique ability of plants to be regenerated from individual cells or tissues. This property, called 'totipotency', means that parts of plants, their tissues or cells, when grown on the right medium in the presence of the right hormones, can be used to regenerate whole plants. (This property explains the ability of some plants to be grown from cuttings or a single leaf.) Thus, by taking plant cells or tissues and subjecting them to transformation, and then regenerating whole plants with selection (to select for transformants in much the same way as that used to select for the presence of an antibiotic resistance gene contained on a plasmid in *E. coli*), it is possible to produce transgenic plants. Because the plants regenerated in this way are derived from one transformed cell, all cells in the regenerated plant (including the germ cells) contain the foreign gene and the plant can then be propagated by the usual means.

If you can transform plant cells and regenerate whole plants containing the foreign gene in every cell, then how do you transfer that foreign gene into plant cells in the first place? As you will recall from Chapter 2, DNA is frequently transformed into *E. coli* by treating the cells with an ice-cold salt solution and then subjecting them to 'heat shock'. However, such a method does not exist to transform plant cells, and other alternatives are employed to make plant cells take up DNA. The two most widely used methods are 'Ti plasmid-mediated transfer' and 'microprojectile bombardment'.

The principle of Ti plasmid-mediated transfer is relatively simple. There is a bacterium called *Agrobacterium tumefaciens*, which carries a plasmid called the Ti (tumour-inducing) plasmid. This bacterium has the ability to transfer the Ti plasmid to plant cells and, once transferred, causes the formation of tumours called crown galls in plants such as grapes. *Agrobacterium* does this because the crown gall is a protective environment in which the bacterium can live and which also supplies nutrients.

Obviously such tumour-inducing activity is undesirable in plants targeted for genetic transformation, so the Ti plasmid has been modified using gene cloning techniques so that it no longer has this ability. At this stage the Ti plasmid is said to be disarmed—the disarming of the Ti plasmid was achieved by cutting out the genes encoding the tumour-inducing activity with restriction enzymes. Therefore, by inserting foreign genes in the disarmed Ti plasmid, placing the recombinant plasmid in *Agrobacterium*, and incubating this bacterium with plant cells or tissues, foreign genes can be transferred into recipient plant cells.

Whilst the employment of disarmed Ti plasmids is effective in transforming plant cells, this method can only be applied to a limited number of plant cell types. One alternative and frequently used method of plant transformation is the use of microprojectile bombardment (also called biolistics). In this process, gold or tungsten particles of 0.4 to 1.2 thousandths of a millimetre in diameter are coated with DNA and shot into plant cells at around 300 to 600 metres per second using special guns employing either gun powder or, more usually, compressed air or helium (Figure 4.1). The trick with this approach is to propel the projectiles at sufficient speed so that they penetrate cell walls and membranes, and to use enough particles to be effective, but not so many as to irreparably damage the plant cell. Biolistics is particularly versatile and can be used on a variety of plant cell preparations and types.

At this point we have briefly looked at the basic

Figure 4.1 Bio-Rad Helios Gene Gun

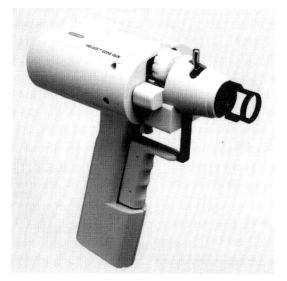

A commercially available gene gun capable of firing particles coated with DNA into plant cells. (Photo courtesy of Bio-Rad Australia.)

principles involved in the production of transgenic plants—transformation of cells or tissue and regeneration of these transformed products into whole plants exploiting the unique totipotency of plants. What we would now like to do is to examine some specific examples of transgenic plants and how they are produced. We hope to give you a taste of how the genetic modification of plants with only one or a few genes can produce quite complex changes in these organisms.

Insecticide resistance and BT-toxin

Our general dependence on crops is the basis of the most obvious goal of plant genetic engineering: the improvement of various traits of crops resulting in increased yields and improved quality. Here is a list of crop plant traits upon which genetic engineering has focused:

- Resistance to pathogens and competitors (weeds, insects, fungi, viruses, bacteria).
- Improvement of nutritive quality.
- Improvement of oilseed quality and quantity.
- Adaptability to physical stress: frost, drought, salinity.

Let's now focus on one of these traits—resistance to insect pathogens. It is of course an understatement to say that the action of insects has a substantial negative effect on the quality and quantity of crop yields. Yearly, world-wide expenditure on pesticides to combat insect infestations of crops and their products is measured in billions of dollars. Moreover, besides a substantial monetary cost, the employment of pesticides also has other negative effects: many insecticides are hazardous and they impact adversely on the environment in general. Moreover, many insecticides will target only a limited range of pests. There are thus clear advantages to be gained by developing insect-resistant crops.

One approach in developing insect-resistant crops employs a suite of proteinaceous toxins produced by the bacterial species *Bacillus thuringiensis* (BT). This microorganism actually consists of a large number of subspecies and strains which produce different toxins (termed BT-toxins) specific for a wide range of target insects. These toxins are specific only for insects, do not persist in the environment and, by way of their mechanism of action, activation and specificity, do not have adverse effects on humans.

The BT-toxins are produced during part of the bacterial life cycle (called sporulation of the bacterium) as large crystalline structures called 'parasporal crystals'. The proteins exist as non-toxic precursors which are converted to a toxic form after ingestion by insects. This conversion of a non-toxic to a toxic form in the gut of insects requires the presence of an alkaline pH and specific protein-degrading enzymes—conditions peculiar enough that it is very unlikely that toxin conversion would occur in non-target species such as humans and animals.

BT-toxins have been used in sprays to control insects—one of these toxins was used in Canada to control the attack on forests by spruce budworm. However, there are limitations to this approach. The toxin is extremely short-lived in the environment and rapidly degrades in the presence of sunlight. Moreover, since many insect pests attack the internal parts of plants, simple topical application of the toxin to the plant as a spray may not be effective. Additionally, the toxin can only kill the insect during a particular developmental stage and thus, considering the short life of the toxins in the environment, timing is critical in the application of the toxin.

One approach to circumvent these problems is to clone the gene or genes encoding BT-toxins into plants. This has been done and, not surprisingly, follows the principles and ideas we have previously covered. Firstly, of course, the gene encoding a particular BT-toxin must be isolated. This involves making a preparation of bacterial DNA employing those processes we encountered in Chapter 1. Cells of the particular strain of BT are grown, harvested and broken open and the DNA purified by chemical methods. The DNA is then digested with a restriction enzyme and ligated into a plasmid. Cells of *E. coli* are then transformed with recombinant plasmids and plated onto a medium that selects for transformants. The trick now of course is to find the colony or colonies containing the gene encoding the toxin. Screening methods such as the use of DNA probes can be used in this process.

After the *E. coli* cell containing the toxin gene has been identified, the gene can be cloned into a disarmed Ti plasmid. The modified Ti plasmid containing the

BT-toxin gene can then be used to transform, for instance, cotton plant cells and tissue which can be regenerated into complete cotton plants. These transgenic plants have been found to be more resistant to some insect infestations than the parental cotton plant that does not contain the BT-toxin gene. In fact, in many cotton-producing countries, transgenic plants expressing the BT-toxin gene have been planted in commercial fields. The advantage to the environment is that these fields do not need to be sprayed as regularly to eradicate insect pests. Thus there is less insecticide in the environment, less expenditure on insecticides by farmers and less collateral damage to other benign insect populations.

Flavr Savr tomatoes

Flavr Savr tomato plants have been genetically engineered to produce fruit that ripens on the vine but stays more robust and less amenable to physical damage, particularly during transport, compared to its non-engineered counterparts. Flavr Savr tomatoes were developed because vine-ripened tomatoes taste better than those not ripened on the vine. But vine-ripened tomatoes are relatively delicate, easily damaged, and overripen quickly. Thus it is difficult to avoid overripening before reaching market and the fruit is therefore costly to transport.

To avoid the problems encountered in using vine-ripened fruit, tomatoes are usually picked when they are green, relatively robust, and have a long storage life. Prior to sale, the fruit is exposed to ethylene gas to develop the red colour, but this process does not produce flavour development. This is the reason why most tomatoes have little taste, and those that do are usually marketed as 'vine-ripened' and cost considerably more. The solution, then, is to produce fruit which can be vine-ripened, but which is physically robust enough to withstand the rigours of transportation. This is exactly the aim in developing the Flavr Savr tomato.

So how were Flavr Savr tomatoes produced? It probably defies imagination how one could engineer a plant that produces fruit with this quality. Clearly it would be silly to say that you just transform plant cells with a 'ripen robustly' gene. To answer this question, one must think about the basic biology which is involved in this process: a gene encodes an enzyme or protein—what does that protein *do* that confers the ability to ripen whilst maintaining a more robust nature?

Firstly, we need to examine what is happening in the ripening process. In part of this process, a large molecule, pectin, which is in part responsible for the structure of plant cell walls, is broken down. This process results in the structural rigidity of the cell wall being diminished, leading to a softening of the fruit. Pectin is broken down by an enzyme called polygalacturonase (PG). The development of Flavr Savr tomatoes has relied on interfering with the production of PG.

How this is done is quite remarkable and to explain it we must first revisit some ideas we encountered in Chapter 1. If you recall, proteins, of which enzymes like PG are a subset, are produced by the process of transcription: genes are transcribed to messenger RNA (mRNA) and then that mRNA is interpreted in the process of translation to produce the proteins themselves. Thus, to interfere with the production of a particular protein such as PG, you could interfere with either transcription or translation. In the case of Flavr Savr tomatoes, it is the process of translation that is interfered with.

To produce Flavr Savr tomatoes, plants containing an antisense version of the PG gene were developed. What this means is that the antisense gene encodes an mRNA which is complementary to that of the PG mRNA. Complementary means that if there is a sequence such as

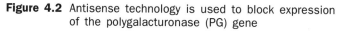

Figure 4.2 Antisense technology is used to block expression of the polygalacturonase (PG) gene

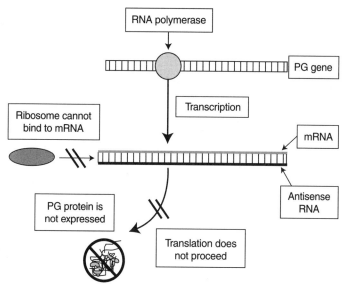

GAUCCGUAG for a particular gene, then its complementary counterpart would be CUAGGCAUC. Complementary sequences pair or bind with one another, so that if a sequence existed in a cell and it were to make contact with its complementary counterpart, a double-stranded molecule could result. Now, because translation involves the binding of a ribosome to the mRNA molecule, if the mRNA is already bound to another complementary molecule, ribosome binding cannot take place and therefore no translation can occur. Thus, by supplying a gene encoding an antisense mRNA to PG, translation of the PG mRNA will not occur and PG protein will not be produced. This was how the Flavr Savr tomato was produced (see Figure 4.2).

Flavr Savr tomatoes have the distinction (dubious or otherwise depending on your point of view) of being the first genetically engineered food to be released onto the US market. The company involved in producing the Flavr Savr tomato, Calgene, naturally enough wanted their product marketed. However, controversy ensued, not because the plants contained a PG antisense gene, but because they also contained a gene encoding antibiotic resistance. If you recall, in most transformation procedures there must be some way of selecting transformed from non-transformed cells. Frequently this employs the use of genes encoding antibiotic resistance. Development of the Flavr Savr tomato was no exception and involved a gene encoding resistance for the antibiotic kanamycin.

Why could there be a potential problem in producing a food containing a gene encoding antibiotic resistance? The answer is that the gene encoding resistance for the antibiotic kanamycin in the Flavr Savr tomatoes may be transferred to bacteria, either in the field where the tomato is grown, or in the gut of humans (or other animals) after ingestion. Such a concern is justifiably significant since the outcome of a transfer of antibiotic resistance genes would be to increase the frequency of antibiotic-resistant bacteria. (Many bacteria are already antibiotic-resistant because of the indiscriminate use of antibiotics, and many of the early antibiotics are virtually useless today.) The point is that any process that may further exacerbate what is potentially one of the greatest problems facing medicine and public health in the next century should be curtailed.

How could antibiotic resistance genes be transferred from transgenic plants to bacteria? The answer is by natural transformation. As transgenic plants are subjected to a range of hostile events (including ingestion) that result in their damage, the cells that comprise them can be broken and the contents, including their DNA, can be released. Such processes would obviously occur in the human gut and also in the environment in general. Bacteria which come into contact with this DNA and which are naturally competent may be able to take up this DNA (including the DNA coding for antibiotic resistance) in a process related to that employed in artificial transformation.

The question then is, 'Will the development and consumption of antibiotic-resistant transgenic plants pose a significant risk in causing the proliferation of antibiotic-resistant bacteria?' In the case of the Flavr Savr tomato, the answer, at least from the perspective of the US Food and Drug Administration (FDA), is that there is little risk—in May 1994, the FDA granted approval for the Flavr Savr tomato to be sold. By doing so, the Flavr Savr tomato made history as the first genetically engineered whole food to gain approval in any country.

The decision of the FDA was not made without considering scientific data. Experimental evidence using animals showed that the enzyme which breaks down the antibiotic was destroyed in the gut. More importantly, there was evidence that it was highly unlikely that the antibiotic resistance gene could be transferred to *E. coli*, which is naturally present in the human gut.

This whole process—the transfer of DNA from very unrelated organisms such as from plants to bacteria—is termed horizontal gene transfer (HGT). HGT contrasts markedly from vertical gene transfer, which is the normal situation, in which DNA is transferred from parent to offspring. In addition to the data concerning the Flavr Savr tomato and HGT, other scientific literature suggests that HGT from plants to prokaryotes is an extremely rare event—the most recently claimed HGT event from plant to bacterium probably occurred over ten million years ago. Moreover, laboratory studies, using optimised conditions, generally report very low or undetectable frequencies of horizontal gene transfer. Thus the literature at this point strongly suggests, but certainly does not prove, that HGT from plants to bacteria is so rare an event that it can be reasonably discounted as a problem when employing the use of transgenic plants.

However, much caution should be applied here: HGT brings up some extremely complex issues, such as gene transfer mechanisms, DNA availability, translocation, DNA stabilisation and, probably most importantly, the role of selection pressure. We urge you not to jump to any conclusions regarding the role of HGT, and suggest that you consult recently published literature if you want to know more.

Plants as biological factories: vaccine production

In the previous examples of using transgenic plants, we were primarily concerned with modifying particular traits of plants to improve their qualities in some way. However, as discussed in the introduction to this chapter, transgenic plants are also potentially very useful as biological factories. One of the main arguments for using transgenic plants in this way is that plants provide a potentially inexpensive alternative to complex fermentation technologies. However, there are other reasons: employing plants as biological factories could minimise the amount of preparation needed before the products are put on the market. For example, if the product is intended to be eaten, it may be possible to produce it in an edible fruit or vegetable, which of course circumvents the need for complex preparative procedures. In addition, some products (such as vaccines) require special storage procedures prior to use. In this case, the employment of plants may also prove beneficial in supplying the products on a user-friendly, as-needed basis.

Now for an example of the employment of plants as biological factories. Probably one of the most potentially fruitful (if you can excuse the pun) avenues uses plants to produce oral vaccines in the edible flesh of those plants.

You will remember from Chapter 3 that a vaccine is a preparation of killed or disabled pathogen (usually a virus or bacterium) which is administered to a person and which then produces an immune response. This response is 'remembered' by the immune system so that subsequent challenges by the same organism elicit a much stronger and specific immune response, which stops that organism from establishing itself. Such a person is then said to be immunised against that pathogen.

One of the first examples of an edible vaccine is against the hepatitis B virus. This virus infects humans primarily through contact with contaminated blood products and body fluids. Carriers of hepatitis B develop liver disease, which can be lifelong and very debilitating. Additionally, in many cases patients suffering such chronic liver disease may eventually develop liver cancer. The gene coding for hepatitis B surface antigen (which can confer protective immunity when incorporated into a vaccine) has been cloned and fully characterised. This gene has already been expressed in yeast cells and purified in order to produce a recombinant injectable vaccine. Clearly, as stated earlier, the disadvantage of using something like yeast is that growing the organism requires the employment of complex fermentation technology. Moreover, in this case, the protein (hepatitis B surface antigen) must then be purified and, furthermore, stored in the cold prior to use. Clearly, there would be some great advantages if it were possible to produce an oral vaccine against a pathogen like hepatitis B in the edible parts of a plant. The hepatitis B surface antigen gene has then been introduced into plants and has been found to be expressed. Human volunteers who have consumed this edible vaccine have been found to mount an immune response against the hepatitis B virus. Hopefully, it will be shown in the not too distant future that people who consume the vaccine in this way not only avoid a painful needle, but are also protected from disease.

Conclusion

There are potential advantages in employing plants as biological factories rather than single-cell systems to produce recombinant proteins and other products: the most obvious of course is that growing plants does not require complex fermentation technology. In addition, in contrast to microbial fermentation systems which normally require complex mixtures of special nutrient-rich material to optimise growth, the nutrient requirements of plants are very modest indeed because of their capacity to produce organic material from atmospheric carbon dioxide using light. Obviously, the potential use of plants in countries which do not have the resources to carry out large-scale fermentations and purifications of proteins cannot be underestimated—and neither can the potential savings in fermentation infrastructure development. The point is that although generally it is easier to genetically engineer single-cell systems, in the long run, because of other considerations, it may be better to genetically modify plants and use these organisms as biological factories.

5 Transgenic animals and cloning 'Dolly the sheep'

In the last two chapters we have described how recombinant proteins have been produced in prokaryotic cells, such as *E. coli*, animal cells and in plants. In this chapter we would like to describe two separate yet related issues: the production of transgenic animals and the 'cloning' of 'Dolly the sheep'. (The meaning of the word 'cloning' changes when you're talking about whole animals rather than single genes, but this will be explained later in the chapter.) The reasons why Dolly is so important will become apparent as you read through this chapter. Firstly, however, we shall look at just what a transgenic animal is and how such animals are produced.

Making a transgenic animal: how and why

A transgenic animal is an animal which contains a foreign gene in each of its cells and, in this sense, is similar to a transgenic plant. However, the construction of transgenic animals is generally very much more difficult than that of transgenic plants (we will explore why this is the case in a moment). If it is so difficult to make a transgenic animal, why would you want to do so? Well, there are two major reasons. Firstly, transgenic animals are used to produce what are known as 'animal models'. These animal models are used for a variety of purposes in medical research; in particular, looking for cures for diseases such as cancer, Alzheimer's disease, cystic fibrosis and muscular dystrophy. We will describe the use of one such animal model below. Secondly, transgenic animals are used to easily produce recombinant proteins of eukaryote origin (most commonly of human origin, to be used for medical purposes). We will also describe an example of this use.

However, let's start by describing how a transgenic animal is produced. If your aim is to produce an animal in which every cell contains a foreign gene (in this case called a 'transgene'), then the first part of the exercise is to clone the gene you are interested in, as outlined in Chapter 2. First you have to isolate the DNA (if it's a eukaryotic gene, then produce cDNA), digest it with a restriction enzyme, ligate the DNA into a cloning vector and transform the recombinant DNA molecules into *E. coli*. Then the gene of interest must be identified, for instance by using an oligonucleotide probe. After the gene has been identified, it is characterised as described in Chapter 2 and the DNA sequence of the gene is determined. Next, the transgene will be cloned into an appropriate expression vector. This vector will contain restriction enzyme sites to insert the gene. A promoter that will allow expression of the transgene in the transgenic animal will be present next to the restriction enzyme sites.

Once this expression vector has been produced it is time to make the transgenic animal. Before we describe how this is done, it is first necessary to think a little about some basic animal reproductive biology. In Chapter 4 we said that, in contrast to unicellular organisms like *E. coli*, the transformation of multicellular organisms like plants and animals presented special difficulties. In a nutshell, this meant that to produce a transgenic multicellular organism, there must be some way in which all cells in that organism can contain the foreign gene. In the case of plants, the solution was to rely on their special ability to regenerate themselves from a few cells or a small amount of tissue (totipotency). Clearly, animals are not totipotent and thus we cannot just take some animal cells, 'transform' them and grow a whole transgenic animal. (The term 'transform' is not strictly correct here because when this word is applied to animal cells it means that they have become cancerous. Instead, the term 'transfection' is used to indicate the process by which an animal cell has taken up foreign DNA.)

So because animals are multicellular and not totipotent, the only way in which a transgenic animal can be made is by introducing genes into the chromosomes of cells near the beginning of conception when only one or a few cells are present. Thus, as the cells divide, the foreign gene replicates along with the original chromosomal material, and all cells contain copies of the foreign genetic material.

The first method that was used to make animals transgenic was to simply take some of the purified expression plasmid DNA and inject it into a fertilised animal egg. If (and this is quite a big if) the expression plasmid DNA is integrated into the genome of the single cell, then the animal that develops from this single fertilised egg would contain a copy of the newly introduced transgene in all its cells. Of course, after fertilisation, the egg, or group of cells that arise from it, will need to be introduced into a 'foster mother' for the animal to be born. Once the animal is born, the presence of the newly introduced transgene must be confirmed. This can be confirmed by using a DNA probe to detect the presence of the transgene in the putative transgenic animal's tissues (for example, from blood) in much the same way as you can detect a cloned gene in an *E. coli* colony.

This approach is naturally more complicated than that used for either plants or unicellular systems and it generally takes much more time and effort to produce transgenic animals: you would need to extract sperm and eggs from the organism, fertilise the eggs in a test tube (*in vitro*), introduce the foreign genes into the embryonic cells *in vitro*, and then introduce the embryo into a receptive female for gestation. At no stage in this multiplicity of steps is there selection for the transgenic embryo and, because none of the steps can be carried out with 100 per cent efficiency, the actual yields of transgenic organisms are normally no greater than 5 per cent of the transfected fertilised eggs.

Thus, you may have to screen hundreds of animals before an animal that contains the transgene is detected. Also, because the transgene integrates randomly into the chromosome, not all of the progeny will properly express the transgene, that is, there may be too much or too little expression for your purposes. This is because the transgene may randomly insert near a chromosomal DNA sequence that can either increase or decrease the level of transgene transcription. Obviously, all of this can be a very time-consuming, expensive and sometimes frustrating business. Another problem is that the product of all this exertion is a single animal with a limited life span. Once you have the transgenic animal you will then have to undertake a breeding program. Because not all offspring of the transgenic animal and its mate will contain the transgene, as animals only receive half of the genetic material from each parent, in each subsequent generation you will have to screen for the presence of the transgene. Once again, this can be a costly and time-consuming business.

Because of the problems involved in producing transgenic animals using the above method, a new method has been developed for producing transgenic mice (mice are excellent models for the investigation of human disease) which helps improve the chance of identifying them with greater efficiency. The expression vector used in this method contains an antibiotic gene such as the neomycin resistance gene, which can be selected for in eukaryotic cells using an antibiotic called G418. Rather than introducing the expression vector into a fertilised mouse egg, the expression vector is introduced into cells called 'pluripotent embryonic stem cells' (these cells can be grown in cell culture on a small scale, which has advantages over large-scale techniques—see Chapter 3). These pluripotent embryonic stem cells are derived from a brown mouse (the mouse colour is important for reasons that will become clear shortly). Once the expression vector has been introduced into the cell line, those cells that have taken up the expression plasmid can be selected for by adding G418 to the medium in which the cell line is growing. After growing the brown mouse cells for a few days in the presence of G418, only those cells that contain the resistance gene will be able to grow. The presence of the transgene can be detected in the cell line by using a DNA probe. Once again the transgene is labelled and the brown mouse cell line DNA is isolated. The presence of the transgene is detected if the DNA probe binds to the cell line DNA.

The next step in this method is to isolate a blastocyst from a white mouse. A blastocyst is a hollow ball of cells which is derived from a single fertilised egg that has undergone several rounds of cell division. In other words, a blastocyst is a very early form of developing embryo. Once several white mouse blastocysts have been isolated, some of the brown mouse cells containing the transgene are injected into them. The blastocysts are then implanted into foster mother mice.

Figure 5.1 Transgenic mice can be produced by injecting a brown mouse cell line containing a transgene into a white mouse blastocyst to form a chimaeric mouse

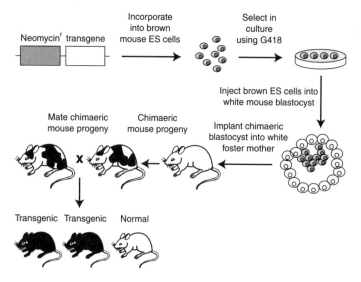

Now you would think that injecting the brown mouse cells into the white mouse blastocyst would upset the development of the blastocyst. However, this is not the case. The brown mouse cells are integrated into the blastocyst and this process results in the birth of a 'chimaeric' mouse, that is, a mouse that contains some cells which were originally derived from a brown mouse (and contain the transgene) and some cells that were derived from a white mouse. Chimaeric mice therefore have patches of white and brown fur. If two chimaeric mice are then bred together, any progeny of this mating that are all-brown in colour will contain the transgene (Figure 5.1). This method of producing transgenic mice requires much less effort because the presence of the transgene is established in the brown cell line through antibiotic selection and can be easily detected as brown mice progeny will be transgenic.

You should recognise, however, that whilst this method of producing transgenic mice improves the efficiency at which such mice can be produced, once the transgenic mouse of interest has been produced, one must still maintain a selective breeding program to keep the transgenic mouse of interest.

A mouse model for Alzheimer's disease

Now that you know it is possible to express a recombinant gene in animals and have an idea about the methodologies involved in doing so, we now need to talk about why you would wish to make an animal transgenic. As we stated in the introduction to this chapter, there are two main reasons for doing so: either to produce a recombinant protein or to use the animal as a model system. In this section, we will examine the use of model systems.

The main rationale for using animal model systems, in particular to investigate human disease, is relatively simple: ethical limitations on animal experimentation are more relaxed in comparison to experiments involving humans. This doesn't mean that one should treat animals in an unethical manner (and most countries have laws against doing so). However, by being allowed some freedom of experimental design when using animals, there is greater scope for success in terms of solving scientific problems. (We realise that experimentation involving animals is in itself an emotive and contentious issue, and it is certainly beyond the scope of this book—and, for that matter, our expertise—to examine the maze of philosophical, technical and scientific issues that arise from this work.)

The example of a transgenic animal model we would like to discuss in this case is the development of a mouse model for Alzheimer's disease. This disease is a progressive neurodegenerative brain disorder which is more likely to occur the older you get. In fact, 30 per cent of people over the age of 80 years are afflicted with Alzheimer's disease. Symptoms of the disease include memory loss, language problems and a progressive physical incapacitation leading eventually to death. As yet, nobody really knows the cause of this disease and there is no cure. Like us, you may have seen the tragic slow and ongoing degeneration in the mental capacity of an older relative that takes place because of this disease, leading to dementia.

A common feature noted in the post-mortem examination of brains from patients with Alzheimer's disease is that of 'plaques' or 'tangles' which are called 'β amyloid plaques'. These plaques are caused by the build-up of a protein called β amyloid protein. It is unclear why β amyloid protein builds up to form plaques, but many scientists are working on the solution to this question and have developed a number of theories as to why this occurs.

So how can the ability to make transgenic mice help to find ways of eliminating β amyloid plaques in the brains of Alzheimer's patients? Well, the gene encoding human β amyloid protein has been cloned, sequenced and characterised. It is possible to take the gene which encodes human β amyloid protein, insert it next to a promoter which is only switched on in the brain, and make a transgenic mouse. This is in fact what has been done. The brains of these mice now produce human β amyloid protein, which has been shown to accumulate there. Thus, by having available the β amyloid protein-producing transgenic mice, it is possible to easily and quickly test thousands of potential cures for Alzheimer's disease using this animal model without having to resort to testing in humans.

This is also a particularly powerful technique because, in the case of Alzheimer's disease, there may be environ-

ɔle in determining the course
sing a mouse model, it is
ɪnder very controlled condi-
ɔnmental factors are involved
ʃ action unthinkable when

Producing clotting factor IX in sheep milk

Haemophilia (Greek for blood-loving) is a disease whereby
the blood is no longer able to form clots. For haemophiliacs, even the most minor of cuts, abrasions and
traumas (for example, small bleeds which occur in the
joints of all of us from time to time) can have severe and
even fatal consequences. It is often necessary following
even minor trauma to transfuse haemophiliacs with large
amounts of blood. There is, however, some light at the
end of the tunnel, as haemophilia can be treated with a
substance which is missing in the afflicted individuals.
However, before we explain what this substance is and
how genetic engineering can be used to improve the lot
of haemophiliacs, you first have to understand something
about how blood clots work. Consider what happens when
someone that doesn't suffer from haemophilia cuts their
finger. As you know, blood will flow out of the wound
for a little while, but not for long. The blood flow will
slow down and a clot will form. Basically, the blood
flowing out of the cut has solidified and thus you will
not bleed to death. Whilst this is certainly advantageous
in terms of stopping small mishaps from turning into
catastrophes, there is also a downside: if the same process
was to occur more or less randomly inside the body, the
results would be extremely nasty. For example, if a clot
forms in the arteries of your heart, you may suffer a heart
attack.

Thus, the formation of blood clots is very tightly

controlled so that clots only form when required. How do clots, which are insoluble, solid structures, form from blood, which is fluid? Well, there are a large number of enzymes that catalyse this process, called 'clotting factors'. (We're going to focus on the enzyme called factor IX.) Probably a good analogy for the large number of enzymes that control blood clotting and are all floating in your blood would be a set of dominoes that are lined up in a fan-shaped arrangement with each domino standing on its end. As each domino falls, it causes two others to fall; thus, progressively larger numbers of dominoes fall over time. When you are cut, it is like pushing the first domino over, this pushes two dominoes over, which then push four and so on, resulting in a cascade of events which leads to a clot.

Haemophilia is a genetic disease, which means that one of the genes that codes for an enzyme involved in the clotting process (such as factor IX) is defective in haemophiliacs. This would be analogous to removing one of the dominoes from the chain—the cascade will start when you cut your finger but it breaks down at the missing domino and therefore a clot will not form. Traditional treatment for haemophiliacs has been to purify the missing factor IX from human blood and continually inject appropriate amounts into the blood-stream, thereby replacing the missing domino.

Since the concentration of factor IX is small, large quantities of blood are needed to produce even small amounts of the factor. You are probably aware that during the start of the AIDS epidemic, many haemophiliacs were infected with HIV because they were exposed to the equivalent of large amounts of blood, and appropriate screening procedures in blood banks were not in place. Although screening procedures for HIV are obviously now in place, it would of course be desirable to have a better supply of factor IX, and of course the answer to this problem is to use recombinant DNA technology.

The human genes coding for proteins that are involved in the clotting process have been cloned and characterised (see Chapter 2), including the factor IX gene. Some of these clotting process genes have been expressed in *E. coli*, which are cultured in fermenters and the gene product purified for use in treating haemophilia. Of course, there is no risk of this recombinant protein being contaminated by human viruses such as HIV or the hepatitis viruses.

Some of these genes have been expressed in eukaryotic cells, grown in fermenters and once again the gene product purified. Again, there is no risk of contamination with HIV or other viruses. However, sometimes eukaryotic genes are not expressed in *E. coli*, sometimes the genes are expressed but the gene product is not functional, or sometimes it is difficult to purify the recombinant protein. This is where transgenic animals can help.

So we may well try to make a transgenic animal which contains the factor IX gene. The question is, where would you wish to have the transgene expressed? It may seem logical at first to produce the human factor IX in blood. However, a moment's further reflection would reveal that this is not the case. Firstly, you would have to continually take blood from the animal. Secondly, you would have to be able to separate the animal factor IX from the human factor IX.

A far better idea would be to produce a protein like human factor IX in the milk of a large animal like a cow or a sheep. The advantages here are obvious: milk is a renewable resource which is produced in substantial quantities and gathered without harming the animal. Moreover, as the foreign protein is produced, it should be modified in ways similar to humans and, just as importantly, this protein should be fairly easy to purify because, whilst milk contains quite a lot of protein, the number of different proteins is limited to about ten different types, thus making it easier to purify the foreign protein. Such a process, of producing a foreign protein for pharmaceutical use in the milk of animals, has sometimes been referred to as 'pharming'.

How is a protein like factor IX expressed specifically in the udder of an animal? In this case, there are promoters which are only switched on in the mammary glands of cattle or sheep. The factor IX gene has, in fact, been inserted next to a promoter called the β-lactoglobulin promoter. This promoter normally regulates the transcription of a gene encoding β-lactoglobulin, a protein found, not surprisingly, in milk. Clearly, following the production of a transgenic animal, all cells in the animal contain a copy of the foreign gene, but it is only expressed in the mammary gland.

In the case of the factor IX gene, it was inserted into a fertilised sheep egg and a transgenic sheep flock developed which produces recombinant human factor IX in the milk (Figure 5.2). Expressing human transgenes in the mammary glands of sheep, goats or cows so far has been shown to have no ill effects on either the transgenic animal or the nursing progeny. It is then simply a matter of milking the transgenic sheep and purifying the human factor IX. The factor IX produced by transgenic sheep is identical in function to factor IX purified from human blood. It is cheap to produce as you only need to feed the sheep grass, and there is no chance of the recombinant factor IX being contaminated by HIV.

Cloning 'Dolly the sheep'

Let us now discuss 'Dolly the sheep'. Up until now, we have used the word 'cloning' to refer to 'cloning a gene'. What this has meant is isolating a gene or genes from one organism and transferring it/them to another organism

Figure 5.2 Transgenic animals can be used to produce recombinant factor IX in milk

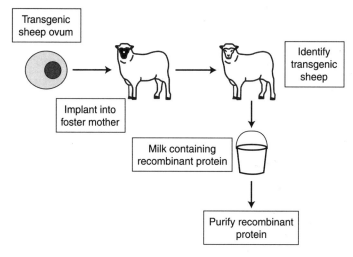

Transgenic sheep ovum

Implant into foster mother

Identify transgenic sheep

Milk containing recombinant protein

Purify recombinant protein

and allowing the gene or genes to replicate in the new organism.

In another sense, cloning means something completely different. The term refers to the ability to produce genetically identical copies of a whole organism. Before we look at what is so special about Dolly, we should first point out that the ability to clone whole organisms, including humans, has to some extent been around for a number of years. This technology relies on the ability to fertilise an egg in the test tube and wait for the fertilised egg (termed a zygote) to undergo rounds of replication (forming two, four, eight cells and so on). At some point relatively early in this process, it is possible to split the clump of cells into fragments, and they will again start to divide as if nothing has happened. These clumps can then be transplanted into the womb of a receptive female and will develop into fully-formed individuals. This technique is called embryo splitting. In this case, since those individuals were derived from one zygote, they are genetically identical and therefore clones.

If you haven't already guessed, a similar process occurs in the natural formation of identical twins, which are in effect clones. In this case, very rarely, after conception the clump of cells will naturally split into two and then develop separately. Just to complete the story, non-identical twins are the result of two separate fertilisation events (involving two eggs produced at the time of ovulation), and thus, they are as different to one another as siblings conceived at different periods in time.

What was the point of the above discussion? Well, one should realise that producing clones of animals is not as technically difficult as would be first thought. However, there is a clear limitation to this type of animal cloning: obviously, one has to be constrained to cloning from cells that exist in early development—it is not possible, using the above technology, to produce a copy of an existing adult (unless you saved, in frozen form, some of the adult's cells from early development). In the rest of this chapter, when we talk of cloning we are referring only to the possibility of cloning copies of an adult individual, using that adult's cells and not using cells from early development.

Up until 1997, the cloning of complete organisms from adult cells, including humans, was pure fantasy. (Hollywood films notwithstanding: in the film *The Boys from Brazil,* starring Gregory Peck, a fanatical scientist clones a number of genetically identical copies of Adolf Hitler from the dictator's cells.) At one time it was thought that cloning mammals from adult cells was impossible, because of innate biological limitations. The reasons for this scepticism were numerous. For instance, in eukaryotic cells many proteins bind to the promoters of many genes blocking the expression of these genes, including some of the genes involved in the development of a fertilised egg into a fully-functioning organism. Another problem is that DNA from mammalian cells is chemically modified by a process called methylation. Getting rid of the methylation so as to allow a cell from an organism to develop like a fertilised egg into another duplicate organism was thought to be impossible. 'Dolly the sheep' changed all that. Let's firstly talk about how Dolly was cloned, then about why Dolly was cloned and, finally, about the ethics of cloning animals and the possibility of cloning humans.

Dolly the sheep was originally a mammary gland cell in a 6-year-old ewe. This and other cells were transferred into cell culture and the amount of nutrient given to the cells reduced. This induced the cells to enter 'quiescence' (a quiet stage in growth and development). The production of quiescent cells was the key to cloning whole mammals. Meanwhile, unfertilised sheep eggs were taken and the nuclei of these cells, which of course contain the genetic material, were removed. These enucleated egg cells were then fused to the quiescent cells and put back into cell culture. Sheep embryos then started to develop. These early embryos, called blastocysts, were then introduced into the reproductive tract of foster mothers. In the experiments that resulted in Dolly, of the 156 blastocysts that were implanted, 21 foster mothers became pregnant. Of these 21 pregnancies, only eight lambs were born (including Dolly), of which one lamb died at birth. Obviously there is still some way to go to optimise the technique.

So the question now is, *why* was Dolly cloned? Well, it was no coincidence that the institute where Dolly was cloned is heavily involved in producing transgenic sheep that express human proteins for therapeutic use. Given the problems with finding a transgenic animal that correctly expresses the transgene, and given the problems associated with having to cross-breed a single transgenic animal to produce a transgenic flock, the ability to clone

Figure 5.3 How Dolly the sheep was cloned

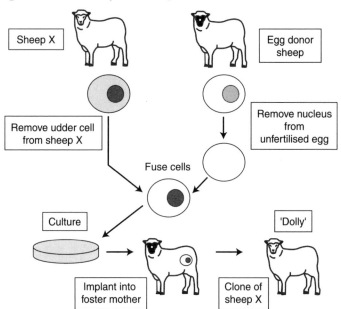

that single transgenic animal and produce a transgenic flock is an attractive proposition. Other potential uses that have been put forward for whole organism cloning are the cloning of prize-winning agricultural stock and the cloning of endangered species.

One can easily see that the ethical implications of whole organism cloning are enormous. In a nutshell, the existence of Dolly opens up an ethical can of worms because this new technology introduces a new element of retrospectivity. Whilst other technologies (such as embryo splitting) allow the production of large numbers of animal clones, there is always an element of uncertainty as to how those organisms will turn out. But with the technology that produced Dolly, one could, for example, make genetically identical copies of race horses that are proven winners. Of course, you could go further to think about the cloning of humans from superior athletes or mathematical geniuses. However, as you are probably aware, genetics is not the sole determinant of an individual's abilities—the environment an individual is brought up in also plays a major role. Here's a list of some other ethical dilemmas which may arise from applying this technology to humans:

- It has been suggested that if a child is lost in a car accident, it could be replaced by cloning. (Because environment plays a big role in development and in personality, the clone may be nothing like the original.)
- Should we bring back people from the past? Albert Einstein, Princess Diana, John F. Kennedy or Martin Luther King? Well, we would need a fresh tissue sample, but maybe in the future even an old tissue sample or a lock of hair might do. Why not clone Jesus Christ (assuming the Shroud of Turin is authentic and still contains intact blood cells)?
- You could produce a clone of yourself. In the event that you need a heart transplant (the defective organ may be the result of environmental rather than genetic causes), you could take your clone's heart, obviously meaning that your clone would die.
- If you make a clone of yourself, which one is the real you: you or your clone?

Conclusion

As you can see, genetic engineering has come a very long way since 1973. Recombinant proteins can now be produced in bacteria, eukaryotic cells, plants and even animals. The proteins that are being produced by this technology are currently being used to treat many of the diseases that afflict humankind. Animal models, primarily in laboratory mice, are also being developed in an attempt to find cures for Alzheimer's disease, cancer, cystic fibrosis, muscular dystrophy and innumerable other human afflictions. Nonetheless, it is important that the general public is well informed about this technology so that we all understand the difference between a transgenic sheep that produces human factor IX to treat haemophilia and a cloned whole organism such as Dolly. With such knowledge, it is then up to the general public and law-makers to decide what is allowable and what isn't with regard to these technologies.

6 Human gene therapy: adapting the Trojan horse strategy

In this final chapter, we would like to describe one of the newest and arguably most controversial aspects of genetic engineering—the genetic modification of humans. Probably the only ethical standpoint from which the genetic manipulation of humans can be easily defended is that of employing the technology to treat human disease. Because of this, the term 'human gene therapy' rather than genetic engineering is used. In this chapter, we will talk only about human gene therapy.

So what is human gene therapy? Well, a simple definition for human gene therapy would be 'the use of genes to treat human disease'. The main aim of human gene therapy is to produce cures for human genetic diseases. Why should this be so controversial, when throughout this book we have given several examples of using recombinant proteins such as human insulin, TPA and GM-CSF (the products of genes) to treat human diseases?

Well, some would suggest that human gene therapy is not much different to making transgenic humans that might be bigger, smarter or more aggressive—something akin to eugenics and constructing the human 'master race'. However, whilst you read this chapter, keep in mind that to make a true transgenic human, you would have to insert DNA into human egg and sperm cells so that the genes you were inserting were inherited. In fact, the development of transgenic humans is specifically prohibited by law in many countries.

What makes human gene therapy so different? In this case, only cells that are not sperm or egg cells (that is, somatic cells rather than germ cells) are genetically modified, and normally only a small fraction of those which comprise the whole person are changed.

We hope that by the end of this chapter you will be in a position to make an informed decision about human gene therapy. Firstly, let's look at a couple of examples of human gene therapy that are currently on trial, and then we will come back to the issue of the ethics of undertaking human gene therapy, once you have a better idea of what it actually is.

A cure for severe combined immunodeficiency?

Severe combined immunodeficiency is an inherited disease, which means a defective gene is passed on from parents to children, resulting in the disease. When a child inherits two copies of the defective gene, this results in the breakdown of the immune system. As the immune system is the major defence mechanism which protects the body from infectious diseases caused by viruses, bacteria and parasites, if it breaks down, you will die from infections that your body normally has no trouble resisting. In fact, this is what happens to people who develop AIDS or 'acquired immune deficiency syndrome' (that is, 'acquired' as opposed to 'inherited'). In this case, it is the human immunodeficiency virus (HIV) that breaks down the immune system. However, let's get back to severe combined immunodeficiency.

Fortunately, this disease is very rare and only afflicts about one person in one million. Perhaps you may have seen the not very good movie (our opinion) about 'a boy in a bubble' starring John Travolta? At the end of the movie, the John Travolta 'bubble boy' character breaks out of his bubble and subsequently dies. If not, then maybe you saw the 'bubble boy' episode of 'Seinfeld' where the bubble is broken by George and the 'bubble boy' gets carried off to hospital? Well, if diagnosed soon enough, babies afflicted with severe combined immunodeficiency syndrome used to be put inside plastic bubbles to live because otherwise, without a functioning immune system, they would die in our environment.

The defective gene that is responsible for severe combined immunodeficiency has been cloned and sequenced. This gene codes for an enzyme called adenosine deaminase. Without this enzyme the cells of the immune system die off and hence the problem fighting infectious diseases—no immune cells means no defences.

So we now know the cause of this disease and the gene that causes it has been cloned. Going back to our definition 'the use of genes to treat human disease', we have the gene which could cure the disease if we can get the normal gene back into the cells of the immune system. Luckily, the cells of the immune system come from the bone marrow, and the techniques used to extract human bone marrow are already in place because they are used in bone marrow transplants. So now we have the disease, the gene which could be the cure if we can get it back into the cells, and the cells themselves.

The problem now is how to get the gene into the bone marrow cells so that it replicates. In this case, inefficient methods of DNA transfer will not be appropriate because as many cells as possible need to contain the gene.

The answer to this problem lies in the esoteric title of this chapter. The biochemical equivalent of a Trojan horse is needed—the piece of DNA containing the adenosine deaminase gene needs to be put into an object that proves so irresistible, it is taken into the cell very efficiently. Moreover, this biochemical Trojan horse must then deposit the adenosine deaminase gene into the chromosome of the bone marrow cells so that the gene is replicated with the rest of the DNA of the cell.

In this case, the biochemical Trojan horse is a virus. What are viruses exactly? These particles are not living (that is, cannot replicate by themselves) and normally consist of DNA or RNA surrounded by a coat of protein, which may also contain fat (lipid). These bizarre things have the ability to infect cells by attaching to them and injecting their DNA or RNA. The nucleic acid encodes proteins (both structural and enzymatic) which take over the normal protein and nucleic acid synthetic machinery of the cell so that it produces more virus particles. The end result of this process, which starts with the attachment of one virus particle, is the release of many hundreds of virus particles, which can then start another round of infection. There are many variations on this general theme of viral infection and replication (and we have already talked a little bit about the diseases caused by viruses such as HIV and hepatitis B), but the above description serves to provide you with some idea of their mode of action. What's important in our discussion here is that some viruses are able to insert genetic information into the human chromosome.

The type of virus that has been used as the Trojan horse to carry the normal adenosine deaminase gene into human bone marrow cells is called a retrovirus. There are many different types of retroviruses. In fact, HIV is a retrovirus. This family of viruses all have the same type of replication cycle. The retroviral genome (genetic information) consists of single-stranded RNA. The RNA is contained within a protein and lipid coat which carries the RNA into the eukaryotic host cell. Once inside, the viral RNA is released and converted into DNA. This DNA then has the capability to insert itself into the eukaryotic cell's chromosome, where the viral DNA 'hibernates' for a period of time. Eventually however, the viral DNA excises from the chromosome and is transcribed and translated, forming new viral particles which leave the host cell through the cell membrane.

Obviously, it is not a good idea to use a fully-functional retrovirus as the Trojan horse because it may be able to cause disease. The scientists who were developing this system also thought about this problem and therefore used a non-functional retrovirus in their studies. A non-functional virus has had some of the genes involved in its life cycle removed using genetic engineering techniques. The loss of these genes means that the virus life cycle does not proceed past inserting the viral DNA into the host chromosome. A neomycin resistance gene was then cloned into the defective virus to enable the selection of cells containing the defective virus using the antibiotic neomycin.

The functional adenosine deaminase gene was cloned into the genome of the non-functional retrovirus. Then, in 1990 at the United States National Institute of Health, two children (aged 4 and 9 years) who suffered from severe combined immunodeficiency were selected for treatment with the Trojan horse virus. A sample of bone marrow was removed and mixed with the virus, which entered some of the bone marrow cells. These cells were then grown briefly in the presence of neomycin so that only those cells which contained the virus and the functional adenosine deaminase gene were able to grow. These cells were then reintroduced back into the bone marrow of the children. It was found that these children started

Figure 6.1 The strategy used to treat severe combined
            immunodeficiency

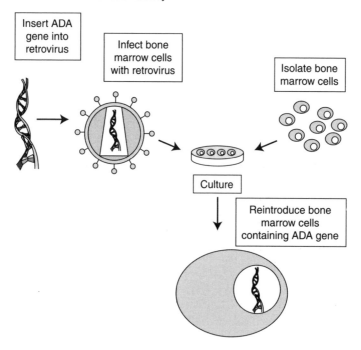

to grow immune cells that produced normal adenosine
deaminase, leading to the development of a functional
immune system (Figure 6.1).

Towards a cure for cystic fibrosis

Cystic fibrosis (CF) is also an inherited disease where a
defective gene is passed on from parents to children
resulting in the disease. When a child inherits two copies
of the defective gene, this results in the occurrence of the
disease. The symptoms of the disease include thick
mucous accumulating in the airways of the lungs, blockage
of the intestine, problems with the pancreas and uncom-

monly salty sweat. It is this last symptom which allows physicians to diagnose CF at an early age. However, it is the thick mucous accumulation in the lungs which is most dangerous. This is because bacteria find this mucous a perfect habitat, which eventually results in terrible cases of pneumonia and eventually death. Twenty or thirty years ago it was uncommon for a CF sufferer to live past the mid-teenage years. Nowadays, with effective antibiotic treatment, CF patients are living into their twenties and thirties and beyond. The problem is, however, that the continuous use of antibiotics selects for antibiotic-resistant strains of bacteria which eventually are able to grow in the lungs of a CF patient anyway.

Unfortunately, CF is a lot more common than severe combined immunodeficiency, especially in Caucasian populations, where the disease has been found to occur in approximately one in every 2500 people. You may well ask why the occurrence of this disease is so high when compared with other genetic diseases such as severe combined immunodeficiency. Given Darwin's theory of natural selection, you would expect that a defective gene with such negative implications for human health would eventually be bred out of the population. Well, in 1998 researchers have shown that one way the bacterium that causes typhoid fever enters human cells is by binding to the product of the normal CF gene. This bacterium is less able to enter cells that express the defective CF gene product. This work suggests that if a human contains one copy of the defective CF gene and one copy of the normal gene, then these people firstly do not suffer CF, and secondly are more resistant to typhoid than those people who contain two copies of the normal gene. Therefore, there may well be a positive implication for human health in bearing one copy of the defective CF gene (resistance to typhoid), which may be the reason for the high occurrence of the defective gene.

Nonetheless, this is cold comfort for CF sufferers (who contain two copies of the defective gene) and their families. The defective gene that is responsible for CF was cloned and sequenced in 1989. This gene codes for an enzyme called cystic fibrosis transmembrane regulator or CFTR for short. This enzyme is found in the human cell membrane and functions to transport chloride across the cell membrane. The defect in CFTR accounts for the salty sweat and the other symptoms of the disease. Amazingly, the most common defect in CFTR results in the loss of only one of the 1480 amino acids that make up the CFTR protein (Figure 6.2).

The cloning of the gene encoding CFTR prompted scientists to think about treating CF using a similar strategy to that used so successfully to treat severe combined immunodeficiency. The question is whether it is possible to use the Trojan virus technique to treat cystic fibrosis. Once again, we have the gene which could cure the disease if we can get the normal gene back into the cells of the lungs. Unfortunately, there are some very difficult technical hurdles which need to be overcome before CF gene therapy becomes a reality. Firstly, it is not possible to remove the lung cells in the same way the bone marrow cells were removed in the severe combined immunodeficiency example above. In this case, if you can't bring Mohammed to the mountain, then you bring the mountain to Mohammed. We thus need a virus that is able to enter the lung cells of a human to deliver the normal CFTR gene. The virus chosen by scientists as the Trojan horse was the virus that causes the common cold, which is called an adenovirus. The adenovirus has a different life cycle to retroviruses. The adenoviral genome consists of double-stranded DNA. The DNA is contained within a protein and lipid coat which carries the DNA into lung cells. Once inside, the viral DNA is released but does not insert itself into the eukaryotic cell chromo-

Figure 6.2 The 1480 amino acid sequence of the cystic fibrosis transmembrane regulator (CFTR) protein given as one-letter amino acid abbreviations (refer to Table 1.1)—the missing amino acid (F), in over 60 per cent of cystic fibrosis patients, is underlined

MQRSPLEKASVVSKLFFSWTRPILRKGYRQRLELSDIYQIPSVDSADNLSEKLEREWDRELASKKNPKLI
NALRRCFFWRFMFYGIFLYLGEVTKAVQPLLLGRIIASYDPDNKEERSIAIYLGIGLCLLFIVRTLLLHPAI
FGLHHIGMQMRIAMFSLIYKKTLKLSSRVLDKISIGQLVSLLSNNLNFDEGLALAHFVWIAPLQVALLM
GLIWELLQASAFCGLGFLIVLALFQAGLGRMMMKYRDQRAGKISERLVITSEMIENIQSVKAYCWEEAM
EKMIENLRQTELKLTRKAAYVRYFNSSAFFFSGFFVVFLSVLPYALIKGIILRKIFTTISFCIVLRMAVTRQ
FPWAVQTWYDSLGAINKIQDFLQKQEYKTLEYNLTTTEVVMENVTAFWEEGFGELFEKAKQNNNNRK
TSNGDDSLFFSNFSLLGTPVLKDINFKIERGQLLAVAGSTGAGKTSLLMMIMGELEPSEGKIKHSGRISFC
SQFSWIMPGTIKENIIEGVSYDEYRYRSVIKACQLEEDISKFAEKDNIVLGEGGITLSGGQRARISLARAVY
KDADLYLLDSPFGYLDVLTEKEIFESCVCKLMANKTRILVTSKMEHLKKADKILILHEGSSYFYGTFSEL
QNLQPDFSSKLMGCDSFDQFSAERRNSILTETLHRFSLEGDAPVSWTETKKQSFKQTGEFGEKRKNSILN
PINSIRFSIVQKTPLQMNGIEEDSDEPLERRLSLVPDSEQGEAILPRISVISTGPTLQARRRQSVLNLMTHSV
NQGQNIHRKTTASTRKVSLAPQANLTELDIYSRRLSQETGLEISEEINEEDLKECFFDDMESIPAVTTWNT
YLRYITVHKSLIFVLIWCLVIFLAEVAASLVVLWLLGNTPLQDKGNSTHSRNNSYAVIITSTSSYYVFYIYV
GVADTLLAMGFFRGLPLVHTLITVSKILHHKMLHSVLQAPMSTLNTLKAGGILNRFSKDIAILDDLLPLTI
FDFIQLLLIVIGAIAVVAVLQPYIFVATVPVIVAFIMLRAYFLQTSQQLKQLESEGRSPIFTHLVTSLKGLWT
LRAFGRQPYFETLFHKALNLHTANWFLYLSTLRWFQMRIEMIFVIFFIAVTFISILTTGEGEGRVGIILTLAM
NIMSTLQWAVNSSIDVDSLMRSVSRVFKFIDMPTEGKPTKSTKPYKNGQLSKVMIIENSHVKKDDIWPSG
GQMTVKDLTAKYTEGGNAILENISFSISPGQRVGLLGRTGSGKSTLLSAFLRLLNTEGEIQIDGVSWDSITL
QQWRKAFGVIPQKVFIFSGTFRKNLDPYEQWSDQEIWKVADEVGLRSVIEQFPGKLDFVLVDGGCVLSH
GHKQLMCLARSVLSKAKILLLDEPSAHLDPVTYQIIRRTLKQAFADCTVILCEHRIEAMLECQQFLVIEEN
KVRQYDSIQKLNERSLFRQAISPSDRVKLFPHRNSSKCKSKPQIAALKEETEEEVQDTRL

some. Instead, the viral DNA is able to replicate and is transcribed and translated in the cytoplasm, where new viral particles form and then leave the host cell through the cell membrane.

Once again, some of the genes involved in viral replication were removed using genetic engineering techniques. The CFTR gene was then cloned into the defective viral DNA. Before undertaking experiments with humans, the defective virus which contained the CFTR gene was firstly administered into rats to see what would happen. As expected, the virus was still able to enter the lung cells and the human CFTR protein was produced.

In 1994, human trials were commenced to determine what would happen when this virus was introduced into the lungs of CF patients.

You could be excused for initially thinking that such a therapy should work, however, nothing is that simple. Firstly, the dose of virus that should be used in humans such that the CFTR gene is sufficiently produced in the lungs has been difficult to calculate. In fact, initial reports from the first human trial suggested that too much virus could induce unwanted side effects (cold-like symptoms) in recipients. Secondly, the inability of the virus to replicate in lung cells would mean that the virus would have to be continually administered. (This is because the virus is not integrated into the chromosome of the lung cells and therefore does not replicate along with the lung cells.) Thirdly, when you catch a cold you do not suffer it for the rest of your days. This is because your immune system wipes out the virus. There really is nothing stopping the immune system of a CF patient from wiping out the defective virus containing the CFTR gene.

From the above case studies it seems that human gene therapy has great potential in the treatment of various genetic diseases but, unfortunately, few human gene therapy projects so far conducted (including the above cases) have succeeded in terms of producing an effective long-term remission of disease. Thus, whilst one can be ever optimistic that one day we may be able to treat the most horrendous of genetic diseases, we clearly have a very long way to go in reaching that goal.

The human genome project

Before we leave this section on human gene therapy, we think it's worthwhile briefly mentioning the human genome project. Whilst this project is not directly concerned with human gene therapy *per se*, the information derived from this project will have such a profound effect on many areas of genetic engineering and molecular biology, it would be remiss of us not to mention it. The primary goal of the human genome project is to determine the nucleotide sequence of the entire human genome. The term 'genome' refers to all the genetic material that an organism contains. This is often measured in base pairs, which refers to the four letter base code (G, A, T and C) which makes up DNA. The term 'base pairs' rather than 'bases' is used to describe the size of DNA because DNA normally exists as a two-stranded molecule where the bases are paired together. To give you some idea of the size of the genomes of particular organisms, bacteria, for example, would have a genome size in the order of a few million base pairs; in the case of a human, this figure is around 3000 million.

Now whilst the goal of the human genome project seems conceptually simple—you just get human DNA and sequence it using well-established technology—the project is a massive technical undertaking: hundreds of collaborating laboratories all around the world are contributing to a project which should take around ten years to complete.

So why would scientists wish to sequence the entire human genome anyway? There are a number of reasons, but perhaps the most important is this: frequently the most difficult part of gene cloning is finding the gene in the first place (as you have seen from the many examples in this book). Just to refresh your memory about how a gene is cloned: you must first extract DNA from an organism, cut it up into fragments using restriction enzymes and clone the fragments into a vector such as a plasmid. You transform cells such as *E. coli* with the plasmids and then screen the colonies for the gene of interest. The screening procedure normally relies on the use of some sort of probe (like a radioactive oligonucleotide probe) or clue as to what that gene encodes in order to screen a plasmid library to find the gene of interest. Such a process becomes very much easier if every human gene has already been sequenced—then, in a sense, all the genes have already been found, and it is much easier to figure out which gene or genes, for example, are involved in genetic diseases such as CF, or diseases such as cancer, or those involved in heart disease or obesity.

Before we leave our brief explanation of the human genome project, we thought we should answer two questions which may have crossed your mind. The first question is, 'Whose genome is it?' This question presupposes that you recognise that all humans, except for those who are identical twins, for example, are all genetically unique. Thus, if one was to sequence the DNA of one human being, it would be similar but not identical to that of another human. So which human being is the subject of this massive project? The answer is that DNA from a variety of humans (or particular cells) is being mapped and sequenced and what results is a database of information from a variety of sources.

The second question is, 'If the human genome project is being carried out by so many laboratories, then why is it taking so long and how is it done?' One major difficulty in the project is that there is a limitation to how many bases can be sequenced in a sequencing reaction. (A sequencing reaction is the main vehicle by which the order of bases in DNA is established and only delivers around 500–1000 bases of sequence information.) Because of the limitation in the number of bases that can be sequenced in a reaction, the fastest way, in the long run, to sequence the human genome is to break it up into small fragments of around 1000 base pairs, clone them in plasmids and then sequence them. In fact, the fragments are produced randomly and are thus overlapping, which means the number of bases that have to be sequenced is even more than the original 3000 million. It doesn't stop there, because sequencing reactions are hardly ever perfect and can deliver spurious results. To counter this problem, the sequencing reactions have to be carried out at least twice to check the results. Finally, after discovering the individual sequences of all the fragments, you would then have to assemble the sequences in the correct order.

This assembly process, in principle, would appear relatively simple: because the fragments derived from the genome are produced by random breakages from a large number of cells, all the fragments would overlap to some extent with their neighbours. One could thus take the sequences and assemble them like a kind of linear jigsaw puzzle. In fact, it's very much more complex than this because the sequence in the human genome is not unique in its entirety, but has substantial regions of repetitive sequence which look like other regions of repetitive sequence. Thus, one can never be sure if a particular fragment of repetitive sequence overlaps with one fragment of repetitive sequence or another.

To counter all of these problems, the human genome has actually been firstly broken down into large fragments of around 200 000–1 500 000 base pairs and cloned into special vectors such as yeast artificial chromosomes (YACs) that can handle pieces of this size. By applying other techniques, it is then possible to determine how the fragments in YACs overlap and produce a map called a 'contig' (contiguous or linear map). One can then take individual YACs and break them into fragments and put them into other types of vectors called cosmids, which accept inserts of around 40 000 base pairs and again, by special techniques, produce a contig map of those cosmids. The cosmids can then be broken into smaller fragments of around 1000 base pairs and cloned into plasmid vectors, from which sequencing reactions can then be carried out. By this process of producing smaller and smaller fragments and making contig maps, one can build up a picture at various levels of resolution and figure out exactly where a particular sequence lies in relation to other sequences. (We left out much detail here, but we hope you get a rough picture of how the project is carried out.)

As you have no doubt realised from the above discussion, this process of contig mapping and sequencing is quite mammoth in scale. In fact, in addition to the project concerned with sequencing the human genome, genome projects on other organisms are also being carried out or have been completed. These projects include, for example, the sequencing of yeast, mouse and various bacterial genomes. Part of the rationale for running other genome projects is that the sequence information for other organisms can be regarded as models for the human genome.

So how far has the human genome project progressed? Well, from what we can gather, around 7–9 per cent of the human genome has now been sequenced, and it is estimated that this task will eventually be completed early in the twenty-first century.

Conclusion

The genetic modification of humans is an obviously ethically contentious issue. However, one must be able to differentiate between the modification of so-called somatic cells in the body and those involved in reproduction (germ cells). Clearly, only the modification of germ cells enables an individual to pass on those changes to their offspring.

Human gene therapy involves the genetic modification of somatic cells in order to treat a particular disease state. So far, a number of trials have been conducted involving human gene therapy, and whilst some of the results offer hints that human gene therapy may be effective in the treatment of some diseases, to date, none of the trials have shown success in terms of long-term effective treatment of particular maladies.

The human genome project is part of the current explosion in genetic information. In the near future, there will be a mass of information about not only individual genes, but whole genomes. This information will enable the rapid location of particular genes of interest and further our understanding of genetics in a profound manner.

Epilogue

You've now come to the end of the book, and have (we hope, if we haven't failed our mission too badly) some understanding of how genetic engineering works. If you recall, we started with a pretty rough definition of genetic engineering—taking DNA from one organism and putting it into another—and have tried to show just what that definition means using some basic ideas in biology concerning the nature of cells, organisms, and the relationship between DNA, RNA and protein.

We hope that this book will serve as some kind of base from which you can further investigate and think about aspects of genetic engineering and molecular biology. An excellent place from which to start would be the journal *Scientific American*, which often carries topical reports on aspects of these issues.

You should also realise that we have limited the scope of this book to deal primarily with the genetic modification of organisms (genetic engineering), but that there is much in regard to the use of recombinant DNA technology that we have left out. The section on the human genome project would have given you some hint that there are other aspects of the use of recombinant DNA technology which, although strictly not involved in the genetic modification of organisms, are nevertheless very important. Other issues which we did not cover include, for example, the use of genetic screening techniques including DNA fingerprinting, which have the potential to impinge greatly on issues of human rights and privacy. To cover such issues and to do them justice would necessitate another whole book.

Clearly, a hallmark of our humanness is our unique ability to be able to substantially alter our environment, ourselves and the living creatures in it. Genetic engineering represents just one newly developed facet of our insatiable appetite for developing and applying technology. It's a hackneyed phrase, but it's true nevertheless, that technology itself is neither good nor evil—it is amoral. As such, we have a responsibility as thinking creatures to assess its use—and abuse—using our unique abilities of reason. Such reasoning can only arise from accurate factual material and premises.

Therein lies the problem: much technology is based on sound principles gleaned from hard scientific work. In the case of genetic engineering, for example, you have seen how an understanding of some ideas in molecular biology was essential to understanding the way in which this technology works. Often those principles can be grasped by the intelligent layperson, but the crunch is that this requires much hard intellectual work and toil.

All we've really tried to do in this book is supply a few facts about biology from which one can then build an understanding of genetic engineering as it stands at the moment. We hope we haven't been too clumsy in the explanation. It is up to you to take the facts on board and apply your full capacities of reason to judge what is and is not ethically conscionable. Good luck!

Index